Honors
Classical
Mechanics

Honors Classical Mechanics

From Special Relativity to Newtonian Mechanics

Henry J. Frisch

Princeton University Press

Princeton & Oxford

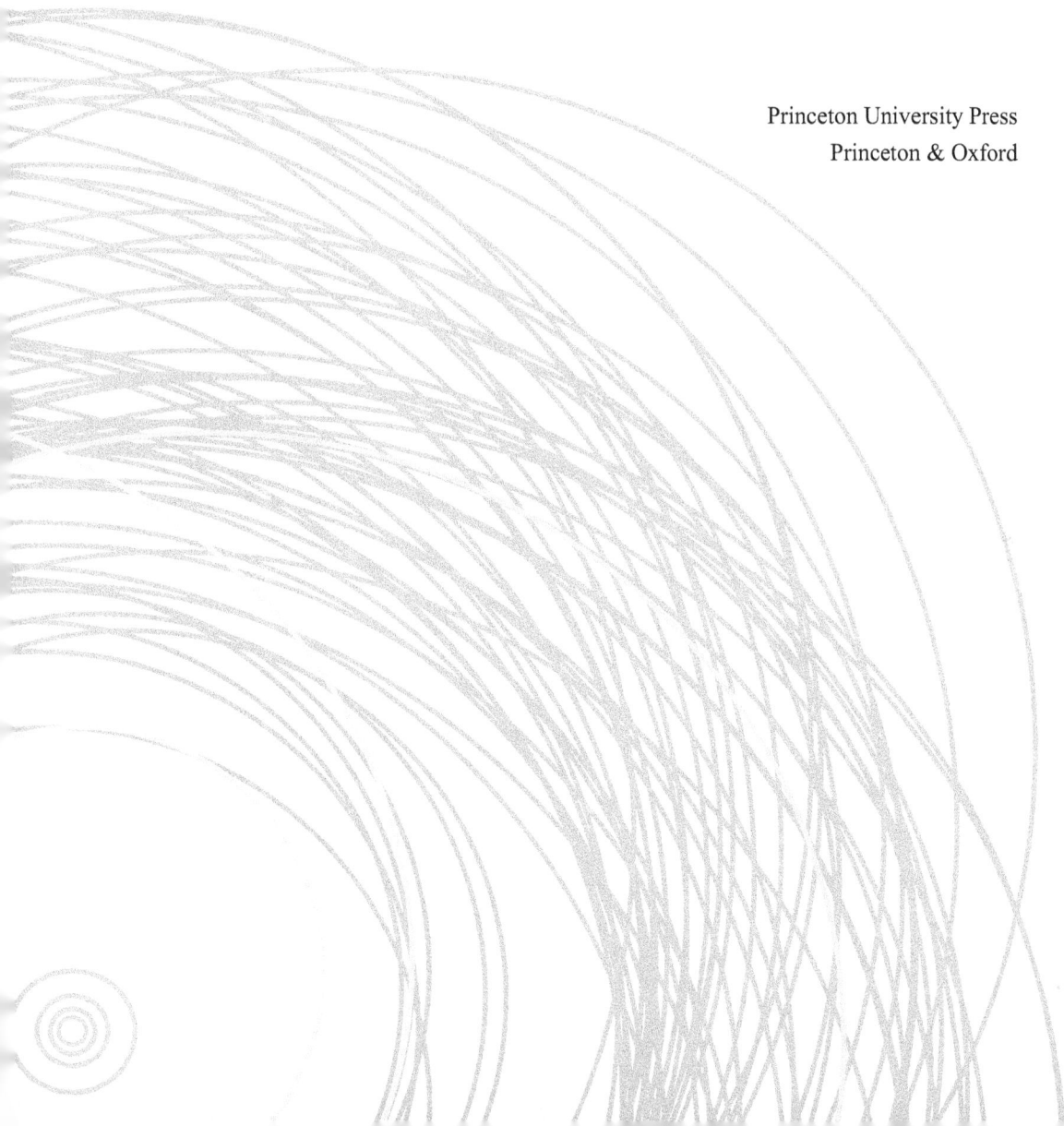

Published by Princeton University Press
41 William Street, Princeton, New Jersey 08540
99 Banbury Road, Oxford OX2 6JX

press.princeton.edu

GPSR Authorized Representative: Easy Access System Europe - Mustamäe tee 50, 10621 Tallinn, Estonia, gpsr.requests@easproject.com

ISBN 9780691277226
ISBN (pbk.) 9780691277233
ISBN (e-book) 9780691277240
Library of Congress Control Number: 2025930824

British Library Cataloging-in-Publication Data is available

Editorial: Abigail Johnson
Production Editorial: Mark Bellis
Text and Cover Design: Wanda España
Production: Erin Suydam
Publicity: William Pagdatoon
Copyeditor: Lor Campbell Gehret

Cover Credit: Courtesy of Scott S. Sheppard / Carnegie Science

This book has been composed in Semplicita Pro and Times New Roman

10 9 8 7 6 5 4 3 2 1

With deep gratitude to my parents, Rose E. Frisch and David H. Frisch,
who dedicated their lives to a deeper understanding of the
natural world for the benefit of humanity.

To Priscilla, Genevieve, and Sarah

CONTENTS

FOR INSTRUCTORS

The following suggestions are an attempt to pass on some of what I have learned teaching this non-traditional curriculum to first-year students more than a dozen times. However, I am sure every class and every instructor will find their own surprises, opportunities, disappointments, and rewards from setting off in new directions.

The Intended Audience: The Broad Category of Honors Students

This book is intended as a one-semester or one-quarter introductory course in mechanics for motivated high school, college, and adult students. First-year college students are expected to have taken a strong high school physics course that covered Newton's Laws, and to have had a mathematics course in single-variable calculus. However, highly motivated students from other backgrounds, with one example being a group of adult life-long learners long out of college, but with some latent mathematical aptitude and a lot of intellectual curiosity, will do well by the end of the course.

The course is meant to be challenging, but enjoyable in an atmosphere of collaborative learning, elimination of competitive behavior, and a high level of strong instructional support.

Rethinking the Goals of the Introductory Honors Physics Course

I have tried to capture several goals of the revised curriculum and course:

1. Developing Confidence; Planning for Success

All students who are in a functioning study group are expected to do well, including building a deep confidence in their own continued questioning of the structure of the physical world. They should be comfortable sharing the tool kits, habits, and language of physicists with each other, the graduate Teaching Assistants (TAs), and the Instructor. The learning environment should be collaborative rather than competitive; grading is a necessary tool, but the goal is to teach rather than rank or weed out. It is important that the students understand that their own grade should not depend on anybody else's grade.

2. Building a Mathematical Tool Kit for Problem Solving

To be able to develop a deep intuition for problems such as rigid-body kinematics and central-force motion, as well as for relativistic kinematics, the students should learn to use a systematic mathematical process rather than depend on their high school physical intuition. Problems such as the Barn and the Pole (Fig. 2.3) can be solved methodically using the mathematically simple recipe of: (1) writing down the 4-vectors for the events in one frame, in this case the four events comprising the front and back ends of the pole reaching the front and back doors of the barn in the runner's frame, respectively; (2) transforming each event into the farmer's frame; and (3) time-ordering the events to reconstruct the history in the latter. In contrast, trying to "wrap one's mind" around the barn and the pole in both frames is uncharted territory.

To this end the instructor should emphasize the importance of the extensive mathematical Appendix, which functions as a parallel text and extended mathematical Glossary. I have found giving equal weight in lecture to the appropriate Appendix section when the need arises is efficient and appreciated.

3. Supporting Broad Intellectual Growth

The curriculum uses precious college time well, covering much of the material of traditional second-year mechanics courses in the first year. The four years of college now support many more requirements than when a traditional physics curriculum was designed, with examples being courses in computer science and molecular engineering, and elective courses in many current topics in physics, astronomy, and cosmology. Most of these courses will be more interesting and relevant to a student's future than the traditional two-course undergraduate Classical Mechanics sequence.

4. Inculcating a Love of Physics

Students enjoy addressing the deep unanswered questions that are typically never asked in an introductory course. A course in 17th Century Mechanics will identify the mathematically-strong students; however, a course with current problems and deep questions may find additional truly excellent students not in the first group, who may also represent a different demographic. Also, the summer after the first year is an ideal time to join a research group; the first year's courses can provide a much more sophisticated basis for participation.

5. Developing the Love of Reading Widely as One Studies

There is a cultivated pleasure in exploring other sources (for example, texts, auto-biographies, biographies, original papers, stories from one's elders) on select topics during a physics course. It is an important habit, and one that is rarely encountered in high school courses or introductory STEM courses. The lists in Recommended Reading are a personal start, and can be supplemented by the instructor's own notes and/or favorite selections.

6. Integrating Young Undergraduates into the Physics Department

As students learn the language, they can attend the Departmental Colloquia, faculty research seminars and group meetings, and can join research groups. My experience is that the Honors students take to this like ducks to water.

Changes in Pedagogical Practices

Changing the goals of the introductory Honors mechanics course requires changes in pedagogical practices. These may include:

1. Treating Problem Sets as Text

The Problem Sets are the primary vehicles in the course for learning. The role is different from that in the traditional 1st-quarter Classical Mechanics class: the problems are meant to be an integral part of the text, to be read along with the chapter, and worked collaboratively in a study group with close contact with the TAs and instructor.

The ability to solve the problems on the Problem Sets with intellectual comfort is the operational pedagogical goal of the course. The weekly Quizzes, the Midterm, and the Final are constructed from the problems on the Problem Sets. Mastery of the problems, demonstrated in closed-book exams with a sheet of formulae provided but no other material, is my goal for the students.

Having high quality solutions promptly available to the students after the Problem Sets are due consequently is essential feedback. Do not settle for less.

2. Employing Grading as Pedagogy instead of Judgment

The weekly Problem Sets and Quizzes are treated as learning tools more than ranking tools. The mantra is, *I am here to teach you and not to grade you. However, grading is an integral part of teaching.*

3. Identifying Inadequately-Prepared Students at the Course Outset

I have found that in a large class usually there are several students who are inadequately prepared for the Honors curriculum. The reasons are diverse: students returning from several years absence, desire for the Honors name on the course for admission to med school or business school, parental pressure, and/or poor advising.

If your institution does not have a targeted placement exam, it is useful to give a brief diagnostic exam at the course start to identify students who do not know single-variable calculus and/or high school mechanics. These students should either have dedicated special help, or should drop the course to take more math or to be in a less demanding sequence.

4. Forming and Sustaining Study Groups

There are a few exceptional students who can do well working alone. However, it is rare—physics is subtle, and working in a study group is essential for all but those few. Identifying students in trouble early will almost always find that they do not have a study group. They should either join a compatible existing group or drop the course (almost all do the former).

5. Teaching the TAs Physics in Parallel with the Students

In these many years I have been lucky to have wonderful advanced graduate students, typically highly-sophisticated advanced theorists, as Teaching Assistants. These remarkable physicists were a key element in the fun.

However if otherwise, be aware that the TAs will likely be learning a great deal of new material along with their students. It is *essential* the TAs use the same notation and conventions. It is very likely they will need to learn entire topics not covered in a typical undergraduate education but covered in the course.

6. Supporting Both Visual and Algebraic Learners

There are two interesting pedagogical aspects of starting with a more sophisticated mathematical language that I do not yet understand fully. The first is that there are students for whom it is natural to learn the algebraic language. However, there are other excellent students who think visually, rely on a limited physical intuition based on an infinite speed of light, and for whom trusting a step-by-step algebraic process is very alien. The visual versus symbolic divide is interesting pedagogically,[1] and should be the subject of discussion with the students while working the problem sets.

---

[1] And is probably changing rapidly with technology.

A second category of student difficulty seems to stem from some students trying to start from the framework of their high school physics courses. Students have been taught the (good) habit of building on what they know, and very good students have become very good at it. Starting far from the Newtonian definitions of energy and momentum is consequently hard for them.

Opportunities

This text is a development of a principle-based version of a first-quarter/semester classical mechanics course. One can imagine exploiting the increased mathematical sophistication of the students at the finish of the course, in particular a comfort with a relativistic tensor notation, for an update to the remaining curriculum. The goals would be to shorten the path to the first-year graduate courses in the main areas, and to provide opportunities for innovative teaching in other areas, both old and new. Electricity and Magnetism is an obvious target; there are other areas, in which there may be local strengths or weaknesses.

I hope this course encourages other teachers to develop a more sophisticated, deeper while also broader, undergraduate physics curriculum. I can attest that it is fun and rewarding.

PREFACE

This book presents a curriculum derived from teaching the Honors Mechanics first-quarter course at the University of Chicago over many years. We start from the principles of translational and Lorentz invariance and develop the necessary language of tensors and multi-variable calculus as we go. The aim is to develop a language capable of addressing fundamental questions on our highly-structured but mysterious Universe.

The book is intended as a long-overdue revision of the traditional curriculum for introductory physics. Honors students are now more sophisticated from their early exposure to scientific research, and the effects of the constant and finite speed of light are now part of their daily lives.[1] Moreover, the course structure of an Honors student's four years now often include substantial time commitments to courses in disciplines not in the typical physics major's undergraduate curriculum of fifty or more years ago.

The course covers the syllabus of the traditional introductory Honors mechanics course. The revision is to start with the principles of invariance under translations in time and space, and their associated conserved quantities of energy and momentum. The treatment assumes a finite and constant speed of light from the beginning. Galilean kinematics is easily recovered by Taylor expansions of the relativistic expressions for energy and momentum after taking the non-relativistic limit $c \to \infty$. A fairly conventional treatment of Newtonian mechanics follows.

My experience from over 50 years of teaching at the undergraduate and graduate levels is that Physics is best learned and taught in its precise native language, which is mathematical.[2] The course introduces tensor notation early, which allows a more concise treatment of both relativistic and non-relativistic dynamics.[3] The mathematical techniques and notation of the chapters are summarized in the Appendix. This serves as a succinct parallel math text, synchronized to the needs of the physics instruction, and with the same conventions and in the same notation. The Appendix is assigned reading in the Problem Sets, which also include math exercises.

---

[1] Location services and GPS in a cell phone, for example.

[2] Consider the content of Maxwell's Equations for electro-magnetism.

[3] And a much leaner, deeper, and more elegant treatment of E&M in a tensor notation in the following quarter.

There are a number of non-traditional ingredients in the revised curriculum and associated learning environment that include: 1) introducing students to the beauty of describing the physical phenomena that arise by taking into account the travel distance of light; 2) "leveling the playing field" by teaching relativistic kinematics in a simplified tensor-based language, new to all students independent of high school and AP preparation; 3) building lasting intellectual friendships among cohorts of students through a non-competitive collaborative study group structure; and 4) enabling early participation of undergraduates in the intellectual life of their Physics Department.

The course relies on a radical change in the traditional competitive learning culture of introductory physics courses. In addition to being much more supportive, working collaboratively is much more efficient than working alone; the inevitable misunderstandings and conceptual questions often are different person-to-person. A like-minded study group can discuss, explain, reassure, and if all are confused, collectively ask for help. Collaborative learning is an essential part of the course.[4]

The course themes are illustrated by comments requested from several students taught with this curriculum in recent years. The responses capture the effect of the revised curriculum and class atmosphere much better than I can.[5]

Student Perspectives

Student 1:

"As an international student, I had some concerns about lacking background when compared to my peers who had had a different High-school curriculum. I remember being in high spirits and ready to try my very best, but this feeling had been laced with some anxiety. [The class] blew all those worries out of the water at the first lecture I attended. . . . The anxiety I held about being behind my peers was all gone, we were all in the same boat. . . . By the time we were done and back to regular programming (pendulums, free-body diagrams, etc.), we all had the capacity to tackle whatever was in the curriculum. All of us in the class, because of what we went through solving problems on SR, [had the] toolset required."

Student 2:

"Before college, I had been taught $F = ma$ in three separate courses from 8th grade through senior year of high school. To see something different was a breath of fresh air. In our problem sets covering special relativity,

---

[4] I now ask a student who does not have a functioning Study Group to find one or leave the course. With rare exceptions, an isolated student is guaranteed not to do well in the course.

[5] Not all students were fans, and there is always *one (not two or none)* thoroughly disgruntled student. I believe him (must be) to have been the same student every year since 1971.

we were tasked to solve the kinematics of different Higgs decay modes.
I felt that in my schooling we had finally covered a modern topic that is
actively being measured. . . . [The class] also gave me the technical tools I
needed for my future courses in physics."

Student 3:

"[That the] course provided the tools to properly understand special rela-
tivity was crucial for my next course—Honors Electricity and Magnetism—
where most of the other students[6] [had] only the typical treatment of 1 week
on special relativity. . . . Since then, the course's material has laid a crucial
foundation for my success in graduate courses—which I started taking less
than a year after this one."

What We Owe a Student in a Revised Introductory Honors Mechanics Curriculum

A first-year student enrolling in Honors Physics has expectations and aspirations. The revised curriculum should provide (at least) the following:

1. Full Coverage of the Traditional Introductory Mechanics Curriculum at Greater Depth

Honors Physics covers all the topics in the syllabus for the introductory first-quarter Honors course in the UC Physics Department, but with Special Relativity at the start of the course rather than at the end. The change is well-motivated, as after several weeks it takes only a few lines to show the foundation of Newtonian kinematics.[7] Students are excited by the beauty of the underlying structures, by the veracity, and by the power of the tools. In addition, the emphasis on collaborative rather than competitive learning leads to a strong cohort of sophisticated inquisitive students that continues in the successive years.

The course avoids the duplication of high school AP courses. Our Honors students often complain that the introductory mechanics course is exceptionally dull, with no new ideas or techniques, and no relevance to any physics questions of interest. Starting with principles of invariance and a non-infinite speed of light is new and exciting, and the more sophisticated language allows an immediate connection to Department Colloquia and faculty research on current questions.

---

[6] This student was in the calculus-based track below the Honors course, which I taught (gently) using this text. The student then moved up into the Honors E&M in the following quarter. The reference to "most of the other students" refers to the continuing Honors students in the class he had just joined.

[7] "You cannot derive a correct theory from a wrong one"—Robert Geroch.

2. The Opportunity for a Deeper but Shorter Undergraduate Intermediate Curriculum; Lightening the Load

A principles-based revision also has a positive impact on the later physics curriculum, as the teaching of tensor notation and a finite speed of light allows for a more sophisticated second-year one-quarter or one-semester mechanics course, with more room for advanced or current topics.

The revised first quarter also allows for a relativistic treatment of Electromagnetism (E&M) in the following quarter. This in turn allows for a more sophisticated intermediate E&M class, with more room for advanced mathematical methods and topics of current faculty research. A revised first-year Honors curriculum will free up time for students starting research early and taking more interesting courses in physics, mathematics, other sciences, and the arts (especially music, it seems).

The upshot is a tight cohort of sophisticated students with a deep appreciation of physics. They will usually go into different fields, but the early introduction to the beauty and power of what Mark Twain described in *Life on the Mississippi* [11] as "such wholesale returns of conjecture out of such a trifling investment of fact" will stay with them all their lives.

3. An Inclusive and Supportive Student Cohort

Starting with a tensor-based treatment of Special Relativity is a "leveling mechanism," in that none of the students have seen it. The perennial problem of "hot-shots" (typically boys from elite high schools competing among themselves) intimidates many students, and is, I believe, one of the factors contributing to the persistent[8] under-representation of women students in the Honors sequence in my Department.

Here I want to quote a student, then a Music major, from a class I taught long ago—not a Physics course, but a Physical Sciences (Gen Ed) course. She writes:

> "This item hit home like a gut punch! It more or less happened to me, but with Chemistry. (I had taken 3 years of chemistry in high school, including Higher Level IB Chemistry, but quickly dropped honors Chem at Chicago because it was just obnoxious.) I believe you're absolutely right on this issue, and not just in terms of gender dynamics, but also factors like geography, public/private school, language, culture, and learning style."

4. Closer Intellectual Contact with the Instructors and Department

However, to successfully teach more material at a deeper level, with higher comprehension and less student stress, requires change in *how* we teach in addition to

---

[8] And completely unnecessary and indefensible.

what. I have learned over many decades that starting from fundamental principles rather than incrementally enhancing high school AP physics requires a new level of close contact with the students. "Relativity" is regarded as fearsome, arcane, and interestingly, not particularly relevant. Part of the awe is historical,[9] and part may stem from the public's confusion of Special Relativity with General Relativity.

Student Outcomes: What Is at Stake?

What is at stake for you, the students, is more than just changing the order of relativistic kinematics and Newtonian kinematics. Learning a language poorly develops habits that once learned are almost impossible to break. The traditional course, often described as an introduction to classical mechanics, reinforces deeply wrong physical intuition. Then as students progress through the physics course sequence, they gradually upgrade their intuition as the language gets more powerful and sophisticated. But, much as a poor accent in a foreign language is deeply innate, a wrong intuition is hard to lose completely.

As one common classical example, the stability of a spinning non-spherical object in free space, stable about two axes but wildly unstable about the third, seems completely un-intuitive.[10] With mastery of the tensor notation introduced here in Chapter 2, one can immediately teach the moment of inertia as a matrix with 6 non-zero components rather than a scalar. Although the resulting 3 stability equations for rotations about the 3 axes are beyond the scope of this one-quarter introductory course, when they arise in the next quarter of mechanics one does not have to unlearn the intuitive (but wrong) mental picture that the angular momentum $\vec{L}$ points along the axis of rotation $\vec{\omega}$. One needs the appropriate mathematical language to concentrate on developing an understanding of the inevitable wobble about one of the axes.

Just as it is not a logical introduction to quantum mechanics, classical mechanics is not a logical introduction to special relativity. My experience is that students find it much easier to start with the correct underlying principles and notation. In the case of quantum mechanics it seems now generally accepted that starting with the mathematics is a better route than the historical one. The intent of this course is to do the same with introductory kinematics.

---

[9] *There were three people named Stein*
There was Gert, there was Ep, there was Ein
Gert's writings were bunk
Ep's statues were junk
And nobody understood Ein.
—Anon (However, see https://quoteinvestigator.com/2020/07/26/ein/)

[10] There is a wonderful video of a spinning T-handle in space linked to YouTube, Reddit, and other sites on the web. The apparent weird behavior is not Nature, but is due to a misinformed human intuition. Honors physics students should be taught that $\vec{L}$ is not a scalar times $\vec{\omega}$.

What Do We Gain?

What do both students and instructors, gain by developing a sophisticated language so early in the curriculum? Let me list the ways:

1. It allows treating both traditional topics and modern areas of research in a unified framework.[11]

2. It is efficient in a student's allocation of courses, covering much of the material of traditional second-year physics courses in the first year. The four years of college now support many more requirements than when a traditional physics curriculum was designed, with examples being courses in computer science and molecular engineering, and elective courses in many current topics in physics, astronomy, and cosmology. Most of these courses will be more interesting and relevant to a student's future than the traditional two-course Classical Mechanics sequence.

3. It allows a more systematic approach to problem solving, typically developed only in intermediate and advanced courses.[12]

4. A Lorentz invariant language allows a sophisticated and efficient treatment of Electricity and Magnetism (E&M) in the following quarter, as E&M is a purely relativistic subject. In one quarter one can cover the material of both the introductory and the following one-quarter intermediate course, and at a much deeper and more satisfying level. It is a beautiful subject.

5. The introduction of vector spaces, projection operators, transformations, and path integrals in Classical Mechanics speeds the teaching of Quantum Mechanics in subsequent courses and disciplines;

6. After even a few weeks we can address unsolved problems in particle physics and astrophysics (for example), with discussions among the students, graduate TAs, and instructor of the many puzzling aspects of our strange and wonderful Universe.[13]

In summary, our understanding of the Universe has changed from the 17th century, the landscape of required skills has changed, and the demography of students has changed. A rethinking of the undergraduate sequence for Physics majors is long overdue.

---

[11] The traditional teaching of Physics was recently described to me as, "The successive telling of smaller and smaller lies." One can ask if this is a necessity or just a family tradition.

[12] On airplanes, when asked what I do for a living, when I reply, "I explore and teach Physics," I invariably get, "Oh! I hated physics!" I also found my introductory undergraduate course unpleasant and unexciting: the exam problems were ad hoc, tricky, and often with an unstated underlying assumption defying common experience (examples on request). We can do better.

[13] The graduate student TAs are interested in a new approach, and also have brought problems and topics from their own work.

ACKNOWLEDGMENTS

First I want to thank Joao Shida, who as an undergraduate and veteran of my course had the time, sophistication, and talent for drawing the Figures while deep in advanced courses and research. Without him the book would not exist. I am also deeply grateful to Mary Heintz, who provided never-failing expert technical support for the manuscript.

Deep thanks to my daughters Sarah and Genevieve and to my wife Priscilla for essential advice and support.

I had gifted teachers, including J.-J. Mayoux (Sorbonne), Eugene Commins (Berkeley), David Jackson (Berkeley), E.E. Moise (Harvard), Lars Ahlfors (Harvard), and Phil Viscuglia (New England Conservatory), who continue to be models of how one can teach ones own thoughts, questions, and understandings to someone else.

The inspiration for starting with the Einstein Gedanken experiments grew out of using *Physics for Poets* by Robert March; (McGraw Hill, 1970) as the text in a Physical Sciences course for humanities students decades ago.

I am grateful to three anonymous reviewers who provided many detailed criticisms; responding to them substantially improved the text. My colleagues Davi Costa, Frank Merritt, James Pilcher, Jonathan Rosner, and Paul Rubinov made trenchant comments on the physics. I am grateful as well to students, past and present, in my group—Evan Angelico, Cameron Poe, and Joao Shida for proof-reading and commenting on the text, and the students from my mechanics course who found typographic errors. I want to especially thank Katerina Steele, who long ago was a student in the Physical Sciences course that led to this approach, and who recently sent an enormously helpful critique of the pedagogical justifications, goals, and prose in the draft *For Instructors* and *Preface* sections.

PART I

Kinematics: Time, Space, Energy and Momentum

The M51 galaxy (right) seen head-on, and its companion galaxy NGC5195, seen edge-on, as they were when the light left it 29 million years ago [1], at the end of the Eocene epoch on Earth. In that time the Sun has traveled more than 1/10 of its orbit around the center of the galaxy. To quote Timothy Ferris in his remarkable book of photographs, Galaxies [2]: "M51 is one of the most achingly beautiful spirals known to humankind, but its beauty contains an element of distress, like that of a racehorse or a bonsai tree. It is in fact a markedly disturbed galaxy. The spiral arm nearer to the companion galaxy reaches beseechingly after it, at one point tucking under an inner arm, while on the opposite side the arm has sprung well away from its normal position." All light travels at a finite speed, and everything we see, far distant galaxies, our Moon in the sky, and even our hands and feet, is as it was at the time the photons were emitted. To think otherwise misses the marvelous depth of time.

CHAPTER 1

Relativistic Kinematics;
Time and Space

Sometime in the early 1970's, Eugene Wigner, one of the architects of Quantum Mechanics and also a very thoughtful philosopher of science [3], was invited to the University of Chicago to give a talk. The conference room, with high windows and ancient stuffed leather chairs used for the weekly faculty-only seminar, was in the old Research Institutes building that housed the Franck and Fermi Institutes. Wigner's talk had been publicly advertised in Chicago; unusually, in the audience there were many old men wearing Eastern European (read Hungarian) suits from the early 20th century.

Wigner gave his talk (I don't remember the topic—it may have been Civil Defense). When he concluded the host asked for questions. From the back of the room an ancient man in an ancient suit asked, "Professor Wigner, do you think we will ever understand it all—that is, we will have a Theory of Everything?" Wigner replied: "Let me tell you a story."

"I had a dog once—a very smart dog. He was so smart: I taught him to beg, to shake hands, to roll over. He learned so quickly that I decided I should teach him to solve Diophantine equations. But you know, it was just beyond that dog."

1.1   Foundations of Classical Mechanics

1.1.1   Introduction: Classical Mechanics as the Limiting Cases of SR and QM

Classical Mechanics gives only an approximate description of motion, as it is a limiting case of each of two *theories*, Special Relativity and Quantum Mechanics (Figure 1.1). Starting with the relativistic expressions for energy and momentum we will derive in just a few lines the classical (approximate) expressions under the assumption that the speed of light is infinite. The classical limit of Quantum Mechanics (QM) is more subtle as QM is non-local in space and time. The limit is

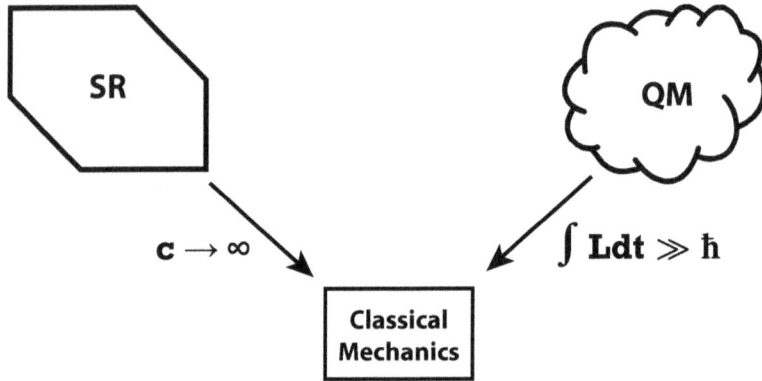

Figure 1.1. Classical Mechanics is approximate, as it is a limiting case of each of two theories, Special Relativity (SR) and Quantum Mechanics (QM). For Special Relativity, the classical approximation corresponds to a world in which the speed of light is infinite, so that we see "everything happening everywhere all at once," i.e., we ignore the time it takes light to arrive from its multiple sources. For Quantum Mechanics, the approximations hold when the energies of a moving object are high enough so that, speaking loosely, effects due to the wave nature of matter integrate to zero along all paths that differ measurably from the classical path.

unfortunately beyond the scope of a one-quarter introductory course, but may be a good subject for an introductory talk in a Discussion Session or office hours after we have learned about path integrals.

1.1.2   Principles of Invariance

Our introduction to classical mechanics rests on three principles. To state them correctly will take developing a small vocabulary for the conditions under which they apply. However, we can loosely state them here now: 1) the laws of physics should be the same for all non-accelerating observers (Lorentz invariance); 2) the laws of physics should be the same at all spatial locations (invariance under translations in space); 3) the laws of physics should be the same at all times (invariance under translations in time). These principles lead to a remarkably concise and elegant description of motion of objects in the classical physical world. Developing this description and a corresponding physical intuition in a concise mathematical language is the subject of the course.

1.2   Inertial Frames

Newton's First Law[1] states that:

All bodies at rest remain at rest or if in uniform linear motion continue in that motion unless compelled to change their state by an applied external force.

---

[1] Corpus omne perseverare in statu suo quiescendi vel movendi uniformiter in directum, nisi quatenus a viribus impressis cogitur statum illum mutare [8].

A reference frame in which Newton's First Law is true is called an Inertial Frame. This may seem like a tautology (true by definition), but isn't. If in a frame any object follows Newton's First Law, then all other objects will also obey the Law, and the frame is inertial.

It is easy to think of counter-examples to Newton's First Law, e.g., rotating frames such as merry-go-rounds, cars going around a curve, or a train accelerating smoothly from a stop. If you have ever tried to walk in a straight line across a merry-go-round in an arcade, you will appreciate George Atwood's First Law of Classical Mechanics:

If asked to work in a non-inertial frame, just say "NO."

1.3   Lorentz Invariance: The Principle of Special Relativity

We state the Principle of Special Relativity as:

The Laws of Physics are the same in all inertial frames, i.e., there are no preferred inertial frames of reference.

Let's take this for now to mean that the mathematical description of motion, for example the equations of motion for a particle, are the same in all inertial frames. The transformation of quantities from one inertial frame to another is called a Lorentz transformation, and invariance under a Lorentz transformation we will take to be a requirement to be a Law of Physics.

1.3.1   The Lorentz Invariance of Electric Charge and the Speed of Light

You may have been taught that the Principle of Relativity is "The speed of light is the same for all observers" [9]. Remarkably, electromagnetic waves (e.g., light), gravitational waves, and massive elementary particles all share the same value of c. We habitually call it "The Speed of Light," but it is a much more general phenomenon than purely electro-magnetic, with c being the ratio of length in space to length in time [6]. We have a highly parochial view of the Universe, limiting it to the very small fraction[2] that interacts electromagnetically [5].

1.4   Einstein's Three Gedanken Experiments

To describe motion of an object we need to answer the question *motion with respect to what?* We define a "Frame of Reference" as the surroundings in which position

---

[2] Known matter is only 5%. More than 98% of that 5% is in the binding energy of nucleons (protons and neutrons) that we cannot observe directly.

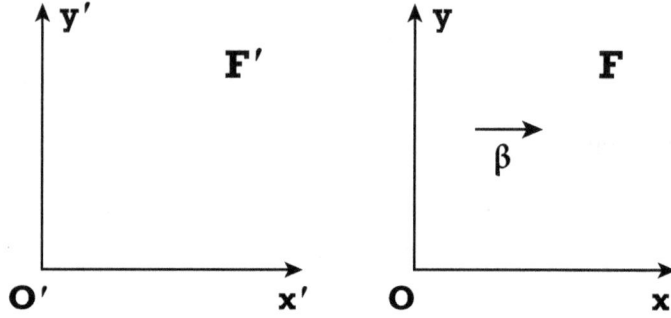

Figure 1.2. The frame convention is that frame F, described by unprimed variables, is moving to the right in frame F', described by primed variables, at velocity $\beta \equiv v/c$. The two origins coincide at $t' = t = 0$. To convert to the alternative convention in which frame F' is moving to the left in frame F, we have only to change β to $-\beta$ in the transformation equations.

and motion are measured. The relative positions of all points in the frame are assumed to be fixed, and are quantified in a coordinate system with an origin, coordinate axes, and scales (the units along an axis). The interior of the train and the station platform in the following Einstein thought experiments below are examples of two frames moving relative to each other (Figure 1.2).

Einstein illustrated the effects of the constant value of the speed of light as measured in different inertial frames through three Gedanken (thought) Experiments [7], each consisting of a train moving at constant speed past a station platform (or vice versa, if you are on the train).[3] Mr. Casals is stationary on the train, which is consequently his frame of reference,[4] denoted as F. Mr. Primrose is standing on the platform near the track; we will denote the platform and station as the primed frame, F'.

We will take the speed of the train in frame F' as $+v$ along the x-axis, i.e., frame F is moving to the right in frame F' at velocity v, and frame F' is moving to the left in frame F, i.e., at velocity $-v$ as shown in Fig. 1.2. For simplicity we assume the two origins in time, t and t', and in space, x and x', of the coordinate systems coincide, i.e., $x' = x = 0$ when $t' = t = 0$.

1.4.1   First Gedanken Experiment: Time Dilation

In the first thought experiment, Casals is traveling on a train. He has constructed a "clock" from a flashing light source on one side of the train and a mirror mounted

---

[3] Einstein is apocryphally said to have asked a train conductor, "Excuse me, could you please tell me when Zurich will arrive?"

[4] Our convention assigns the unprimed frame to the untransformed system in which the initial events occur, and the prime to the transformed frame. To use the opposite convention, replace β with $-\beta$ everywhere. Note that the Lorentz factor γ, which will be introduced shortly in Section 1.4.1, is unchanged. For rotations, the convention corresponds to rotating the coordinate system rather than rotating the vector.

Frame F

Mirror

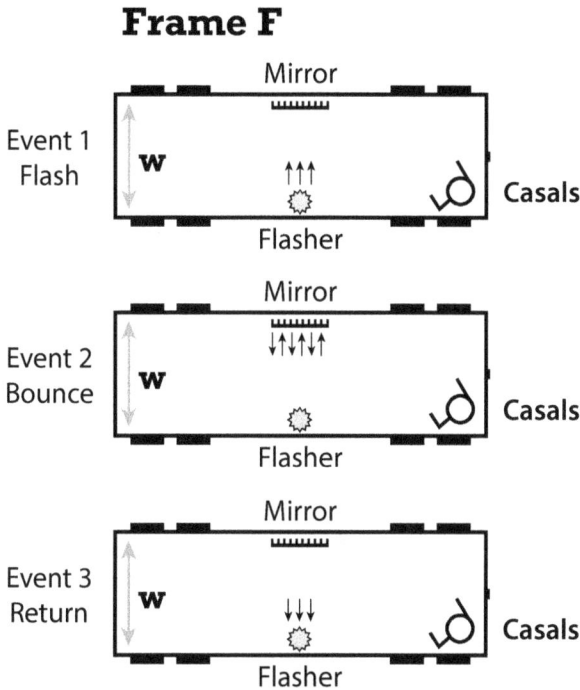

Figure 1.3. Gedanken Experiment 1, the measurement of time, as recorded by Casals. In his reference frame, F, a tick of the clock corresponds to a cycle of 3 events, shown in the 3 panels: (1) A flash of light at the source; (2) the bounce of the light at the mirror; and (3) the return of the light to the source, initiating another flash. The time between ticks of the clock is twice the light travel time across the train.

directly across the train from the light, as shown in Figure 1.3. A "tick" of the clock corresponds to a cycle of three events: (1) The source flashes a short pulse of light; (2) the light bounces off the mirror on the other side of the train; and (3) the light initiates another flash at the source after returning across the train. The clock continuously cycles, with Casals observing the flashes of the clock as his basis for measuring time.

The time it takes light to travel across the train is the width of the train, w, divided by the velocity of light, c. To go and return takes twice that, so the period (time between ticks) of the clock according to Casals is t = 2w/c.

Primrose, however, has a different story, as shown in Figure 1.4. He agrees that the sequence starts when the light flashes. However, the train is moving at velocity v, and so when the light arrives at the mirror, the mirror is not directly across from the light, but has moved along the x'-axis by a distance $x' = vt'$. The distance traveled by the light consequently is longer than w/c. Since light travels at the same velocity in Primrose's frame as in Casals' frame, the time between ticks will be longer, i.e., Primrose records the clock running slower than does Casals. Time is "dilated" in his frame—the separation in time between events in frame F, recorded by Casals, is recorded as longer in frame F' by Primrose.

Frame F'

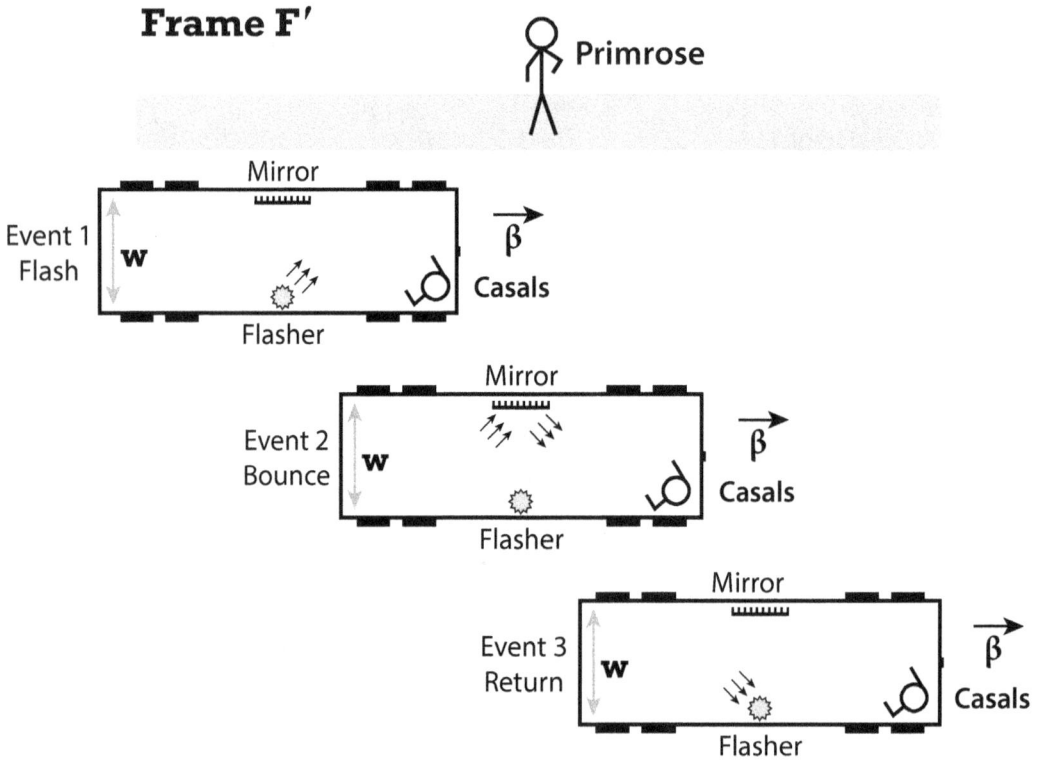

Figure 1.4. Gedanken Experiment 1 as seen in Primrose's frame on the station platform as the train goes by. Primrose records the same set of events as Casals: Event 1 is a flash; Event 2 is the bounce from the mirror; and Event 3 is another flash when the light returns to the source. However, the events occur at different places and times in Primrose's frame than in Casals' frame, in which for example, Events 1 and 3 occur at the same space point.

Casals and Primrose agree that there is a sequence of periodic flashes, but they differ on where and when in their respective frames the flashes occur. We can make this quantitative as follows.

In Figure 1.2 we defined two frames of reference each with a coordinate system: Casals observes life (lives!) in frame F, and Primrose observes life in frame F'. For each of them we will define an "event" as a point in time and space in their reference frame, concisely written as 4 numbers. For example, each flash is an event. In Casals' frame, F is unprimed, and hence a flash has unprimed coordinates (t, x, y, z). In Primrose's frame, a flash occurs at (t', x', y', z').

Figure 1.5 shows the "clock" made from the flasher and mirror in Primrose's frame F'. Let's define t' to be the time it takes the flash to get to the mirror.[5] The distance traveled by the light is ct'. Applying Pythagoras' Theorem to the right triangle (Fig. 1.5) formed by the train width w as one side, the distance ct' traveled by the light at velocity c in the time t' as the hypotenuse, and the distance vt' traveled

---

[5] The time between ticks of Primrose's clock will be twice this, as the light has to go across and come back.

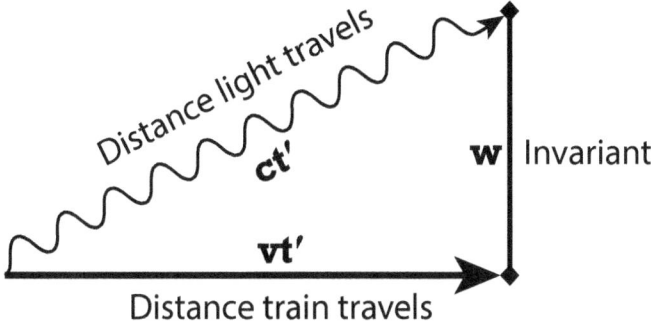

Figure 1.5. The right triangle in Primrose's frame formed by the train width w as one side, the distance ct' traveled by the light in the time t' it takes to get to the mirror as the hypotenuse, and the distance vt' traveled by the train along the x-axis in the time t'. By Pythagoras's Theorem (see Eq. 1.1), the time between flashes seen in Primrose's frame is longer than that in Casals' frame by the Lorentz factor $\gamma = \frac{1}{\sqrt{(1-(v/c)^2)}}$.

by the train in the time t' as the other side.

$$(ct')^2 = w^2 + (vt')^2$$

$$(ct')^2 - (vt')^2 = w^2$$

$$(t')^2 (c^2 - v^2) = w^2$$

$$(t')^2 = \frac{w^2}{(c^2 - v^2)}$$

$$(t')^2 = \frac{(w/c)^2}{(1 - (v/c)^2)} \tag{1.1}$$

$$t' = \frac{(w/c)}{\sqrt{1 - (v/c)^2}}$$

$$t' = \gamma \frac{w}{c}$$

And thus Time Dilation $t' = \gamma t$

where γ (gamma) is the Lorentz factor,

$$\gamma \equiv \frac{1}{\sqrt{1 - (v/c)^2}}. \tag{1.2}$$

Primrose sees the 'clock' of regular flashes running more slowly (e.g., a second takes *longer*) than Casals does by the factor of γ. This effect is traditionally referred to as Time Dilation; time intervals in the moving frame F are measured as "dilated," i.e., lengthened in the laboratory frame F'.

The Lorentz factor γ corresponds to the ratio of the hypotenuse to the side transverse to the direction of motion in the Pythagorean triangle of Figure 1.5. Since the hypotenuse cannot be shorter than a side, the Lorentz factor γ is always equal to or greater than 1. Because the ratio is to the transverse side, which is invariant, the Lorentz factor is (in principle) unbounded from above.

Note that if the speed of light were infinite as it is in the non-relativistic approximation, both Casals and Primrose would measure the time between ticks of their clocks as zero, i.e., all of the events would be simultaneous.[6]

1.4.2   Second Gedanken Experiment: Lorentz Contraction

In the second Gedanken Experiment, the train is again moving at velocity v relative to a station platform. However this time the light travels the length of the train from the flasher at the rear to the mirror at the front, where it is reflected back to the flasher, as shown in Fig. 1.6. On arrival of the light at the flasher, the cycle repeats, making a "clock" as in the first Gedanken Experiment. However, the difference is that the light travels along the direction of motion of the train rather than transverse to it. Casals is on the train[7] and measures the time between flashes and the length of the train L between the flasher and the mirror. As before we call Casals' frame F.

Primrose is on the station platform as shown in Fig. 1.7; we call this the lab (for laboratory) frame, and denote it by F'. We will calculate the length of the train, L', measured by Primrose in F' relative to the length L measured in frame F by Casals.

The first step is to calculate the time in both frames. In F, the time from the flasher to the mirror is $t_{out} = L/c$. The time back from the mirror to the flasher is $t_{back} = L/c$. The total time for one cycle is $t_{Tot} = 2L/c$.

As seen by Primrose in F', however, the train moves as the light is traveling, and so the distance from the flasher to the mirror is longer. Similarly the distance back from the mirror to the flasher will be shorter. Calculating the total time as the sum of the time to go out and the time to come back:

$$\text{Time out}: \qquad t'_o = (L' + vt'_o)/c$$
$$\text{Time back}: \qquad t'_b = (L' - vt'_b)/c$$

$$\text{Solving for } t'_o: \quad ct'_o = (L' + vt'_o)$$
$$(c - v)t'_o = L'$$
$$t'_o = \frac{L'}{(c - v)}$$

---

[6] If this seems obvious, you have already acquired a relativistic intuition.

[7] It doesn't matter where Casals is in the train, as long as he is not moving in it.

Frame F

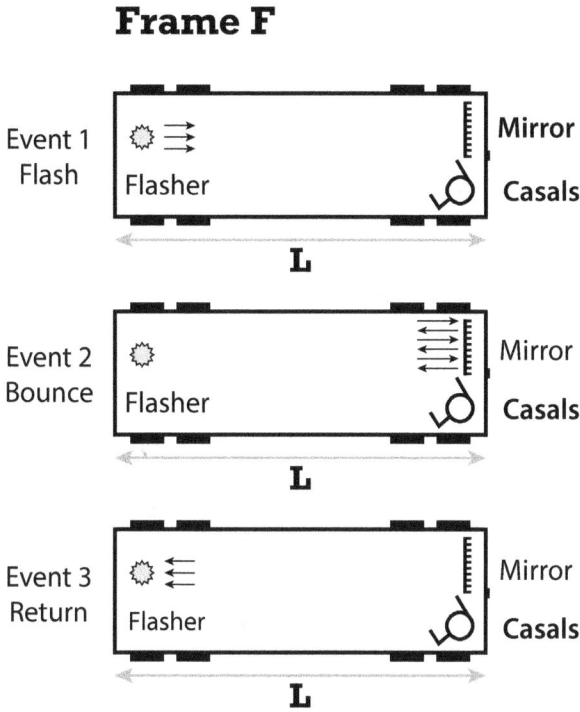

Figure 1.6. Gedanken Experiment 2, as recorded by Casals in his reference frame F. In Gedanken 2 the light travels down-and-back along the direction of motion of the train rather than across it. The 3 events that make up one cycle of the "clock," shown in the 3 panels, are: (1) a flash of light at the source; (2) the bounce of the light at the mirror; and (3) the return of the light to the source, initiating another flash. In frame F, the events occur at fixed locations, i.e., the flasher and the mirror are not moving. The time between flashes is twice the time it takes light to travel the length of the train between the source and the mirror.

$$\text{Solving for } t'_b: \quad ct'_b = (L' - vt'_b)$$

$$(c + v)t'_b = L'$$

$$t'_b = \frac{L'}{(c + v)} \tag{1.3}$$

$$\text{Total time}: \quad t'_T = \frac{L'}{(c - v)} + \frac{L'}{(c + v)}$$

$$= \frac{L'[(c + v) + (c - v)]}{(c^2 - v^2)}$$

$$= \frac{2L'c}{(c^2 - v^2)}$$

$$= \frac{2L'c}{c^2(1 - v^2/c^2)}$$

Frame F′

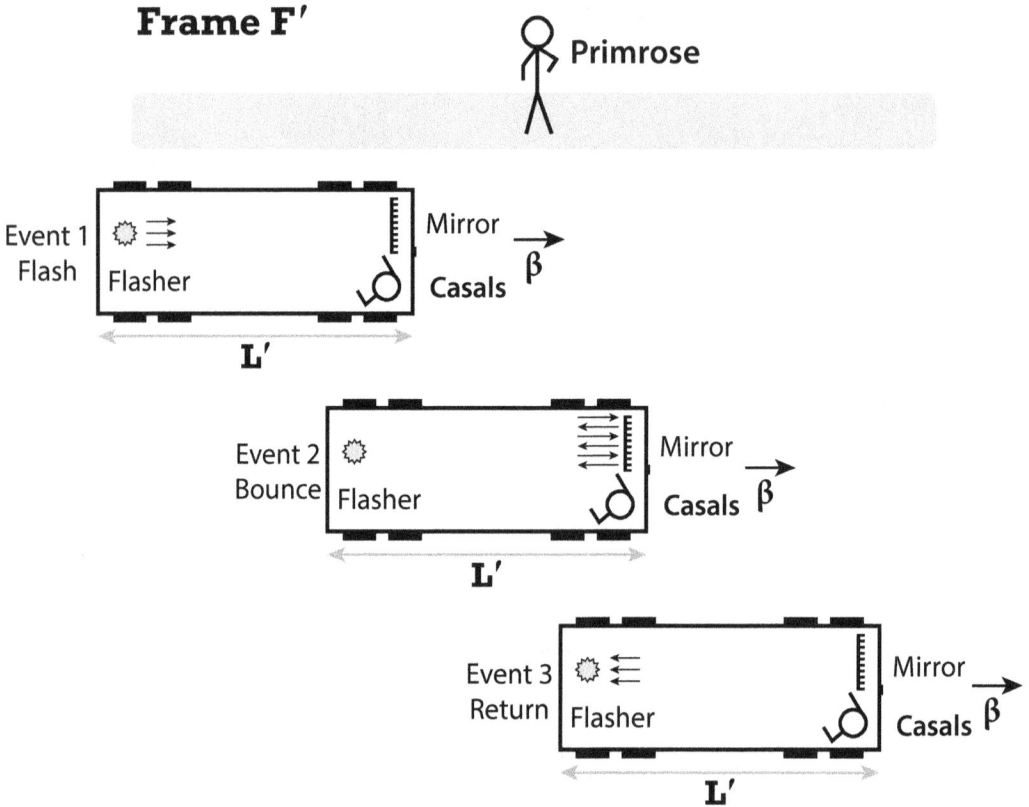

Figure 1.7. Gedanken Experiment 2 as recorded in frame F' by Primrose on the station platform. Top image, Event 1: the source emits a light flash in the direction of the train's motion toward the mirror, which is moving away from it. Middle image, Event 2: the light flash bounces off the mirror, which has moved away during the light transit time, and heads back toward the source, which is moving toward it. Bottom image, Event 3: the flash returns to the flasher, which has moved toward it. Both the times *and* space positions of the events recorded by Primrose in F' differ from those recorded by Casals in F.

$$= \frac{2L'}{c(1 - v^2/c^2)}$$

$$t'_T = \frac{2L'}{c}\gamma^2$$

Solving for L' in terms of t'_T: $L' = \frac{c}{2\gamma^2}t'_T$

where γ is the Lorentz factor.

However,[8] from Gedanken 1 we found that the time interval in F' was a factor of γ larger than in F:

---

[8] This is the final step that I always have to go back to my notes to remember.

$$t'_T = \gamma t_T = \gamma (2L/c).$$

Substituting t'_T into Eq.1.3 : $L' = \dfrac{c}{2\gamma^2}t'_T = \dfrac{c}{2\gamma^2}\gamma(2L/c)$ (1.4)

And thus Lorentz Contraction : $L' = L/\gamma$.

The length of the train in F' is shorter than in F by a factor of γ. This is Lorentz Contraction. Time intervals in F' are longer by a factor of γ (Time Dilation); space intervals along the direction of motion are shorter by a factor of γ (Lorentz Contraction).

What about lengths measured perpendicular to the direction of motion? These are invariant (unchanged) under the transformation between frames. Why are these different from lengths along the direction of motion? In the direction of motion *both* the times and positions of the end-points of the space interval between the two events on the moving train (in this case a light flash and a bounce) are different in the two frames. For the transverse measurement, Casals and Primrose can arrange a simultaneous measurement of the width of the train by, for example, placing metal electrical contacts with one pair jutting out from the train and the other pair on fixed posts on both sides of the tracks, respectively. The two sets of contacts will make contact at the same time and place. They can measure the distance between the contacts in their own frame, and so will agree on the width of the train.

More dramatically, there's a proof by contradiction. Suppose transverse dimensions are shrunk by a factor of γ. Consider a train approaching a tunnel that is narrower than the train. Casals isn't worried, as he says, "Not a problem— I'll slow down and the tunnel will get wider." Primrose, on the other hand, is jumping up and down shouting, "Pablo, for Heaven's sake speed up! The train needs to be narrower!" Whether or not the train crashes would definitively identify either F' or F as a preferred frame. Happily, lengths transverse to the motion are invariant.

1.4.3   Third Gedanken Experiment: Simultaneity Is Not Lorentz Invariant

The third of Einstein's Gedanken Experiments demonstrates that two events measured as being simultaneous in a frame F are typically not simultaneous in a frame F' moving with respect to F, due to the speed of light being finite and equal in the two frames. Because of the motion of F' with respect to F, the simultaneous arrival of light from two events in F requires the two events to be an equal distance from Casals. However Primrose has a different story as we will discuss below.

We have set up the experiment here, but have assigned the solution to the Problem Set. Figure 1.8 shows the same moving train and platform as in the previous two Gedanken Experiments. Casals is now in the center of the train, standing at the window on the side closest to the platform. There are two flashers, one at each end of the train. The flashers are synchronized by Casals so that they flash simultaneously; the flash from each consequently simultaneously arrives at Casals at time $t = L/2c$.

Primrose is standing on the edge of the platform so that his head and Casals' head are very close at the moment when the flashes arrive. Consequently Primrose also sees two simultaneous flashes, one from the rear of the train and one from the front, at the same time as Casals. However, his story is that since the train was moving while the flashes were traveling, to arrive at the same time the flash at the rear of the train had to be earlier than the flash at the front, since the rear was further away and the front closer to the place where he and Casals both saw both flashes. The two flashes are simultaneous in frame F and *not* simultaneous in F'.

1.4.4   The Velocity β and Lorentz Factor γ; Identities and Generalizing to 3-Dimensions

The velocity β and Lorentz factor γ occur in so many situations that it is useful to write out the identities between them so that if given β one can find γ and vice versa. We also give the approximation for β in the limit of large γ (See Section A.1.3 of Appendix A).

$$\beta \equiv v/c$$

$$\gamma \equiv \frac{1}{\sqrt{1 - \beta^2}}$$

$$\gamma^2 = \frac{1}{1 - \beta^2} \qquad\qquad (1.5)$$

$$\beta^2 = \frac{\gamma^2 - 1}{\gamma^2}$$

$$\text{For } \gamma >> 1, \quad \beta \approx 1 - \frac{1}{2\gamma^2}.$$

In 3 dimensions, each of x, y, and z have their respective velocities $\beta_x, \beta_y, \beta_z$, and consequently their respective Lorentz factors $\gamma_x, \gamma_y, \gamma_z$. We will largely stick with uniform motion in one dimension, which always can be taken as along the x-axis.

1.5   A Coordinate System that Accounts for Light Travel Time

Einstein constructed his three Gedanken Experiments as brilliant pedagogy for the lay public. However, we will now leave them to develop a precise mathematical

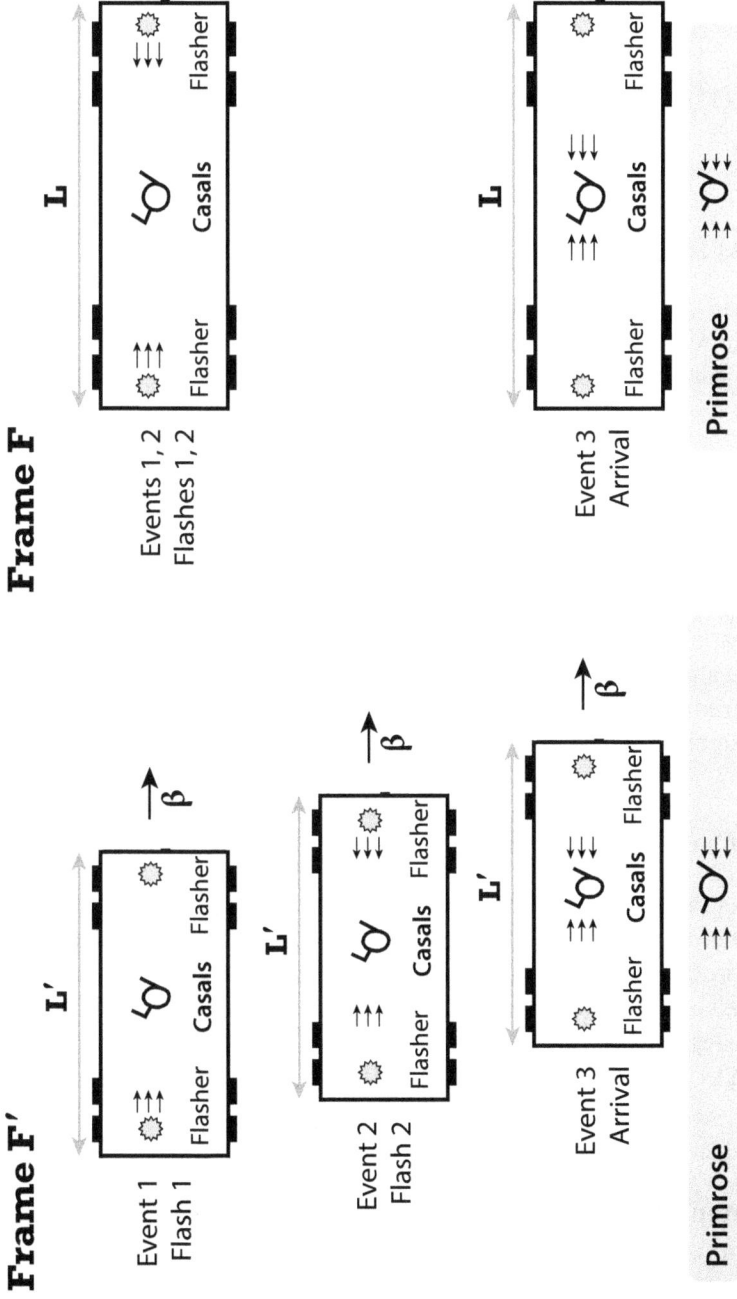

Figure 1.8. Gedanken Experiment 3. Right: Events recorded in frame F. Casals, who is on the train, puts a light at each end of the train, set to flash simultaneously. He verifies this by standing in the middle of the train and seeing the flashes arrive at the same time. Left: Events recorded in frame F'. Primrose is standing on the platform close to the tracks and sees the same two flashes simultaneously at the moment that Casals is directly opposite him. However, Primrose knows he has to take into account the distance the train traveled while the light was propagating. The light from the rear end of the train had to travel further than the light from the front end to arrive at the same time. He concludes that the rear light flashed before the front light, i.e., the two flashes were not simultaneous.

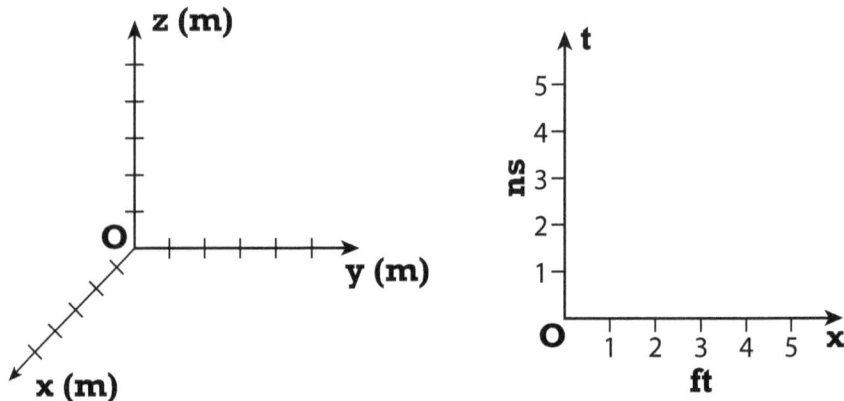

Figure 1.9. The conventions for Cartesian coordinates. Left: in 3-dimensions, a right-handed system with the positive z-axis up, x out of the page, and y to the right. Right: The 2-dimensional xt plane. In both panels note the labeling of the origin, axes, scales, and units.

language that allows us to transform any event, i.e., a time and location, from one frame into another frame.

Once we have the events in the new frame it is straightforward to calculate the intervals in space and time between one event and another event. We will find that there are invariant lengths, numerically the same in both frames. Wonderfully, the same mathematical framework applies directly to the transformation of energy and momentum from one frame to another, and to the transformations for electric charge and the Coulomb potential of classical Electromagnetism.

1.5.1   A Cartesian Coordinate System in 3-Space

Kinematics is the science of motion in the absence of applied force. To be quantitative, we need a coordinate system consisting of axes that span the space (i.e., every point in the space can be reached), and scales on the axes that provide a numerical value of the position along each axis. We will work initially in Cartesian coordinates, with three orthogonal axes. Figure 1.9 shows our conventions: a right-handed system[9] with the positive z-axis up; x out of the page; and y to the right.

1.5.2   Extending the 3D Coordinate System to Account for Light Travel Time

We have seen in the three Einstein Gedanken experiments that both the time and the place of events are needed to characterize what is seen in a given frame. In

---

[9] See Section A.1.1.3 of Appendix A for a definition of "right-handed" in vector notation.

3-dimensions vector notation provides a remarkably powerful calculational framework.[10] The fixed ratio of travel time to distance in every inertial frame, which we denote by the letter c, allows the incorporation of time into a dimensionally-correct vector framework.

1.5.3 The 4-Vector $x^\mu = (ct, x, y, z)$

We will define the time and place of an event in a given frame by a "4-vector" x^μ:

$$x^\mu = (ct, x, y, z) \tag{1.6}$$

where μ is an index[11] that runs from 0 to 3, with x^0 being ct, and x^i being x,y,z for $i = 1, 2, 3$, respectively.[12]

1.5.4 The Invariant Length of a 4-Vector

We saw in the first Einstein Gedanken experiment that the times and places of the events—the flashes of light and the bounce from the mirror—were different for Casals and Primrose. However, we can define a "distance" in 4-dimensions that they agree on, i.e., is invariant under the transformation from one frame to another. We call this the "invariant length" of the 4-vector.[13] It corresponds to the side transverse to the direction of motion; both Casals and Primrose agree on the length of the side. The relationship of the sides of the triangle in Figure 1.5 is given by the Pythagorean Theorem: the square of the hypotenuse is equal to the sum of the squares of the two sides. The square of the transverse side is thus the hypotenuse squared minus the square of the side in the direction of travel. Referring to Figure 1.5, in 4-vector notation, both Casals and Primrose agree on the invariant length squared:

$$|x^\mu|^2 = (ct)^2 - x^2. \tag{1.7}$$

More generally,

$$|x^\mu|^2 = (ct)^2 - x^2 - y^2 - z^2. \tag{1.8}$$

The invariant length formally is the square root of this; however, square roots are a pain for a quantity that can be positive or negative, and so I usually refer to the square as the invariant length, with the understanding that to get the correct numerical value and units one should take the square root.

---

[10] For examples see Section A.1.1 of Appendix A.

[11] For practice with indices see Problem 2.

[12] There are many conventions for 4-vectors; the choice doesn't matter as long as one is consistent. Another convention we are using includes using Greek letters for 4-vectors (μ is the Greek "m" (pronounced *myou* rather than the bovine *moo*) and Roman letters for 3-vectors, for example i, j, and k. If you don't know the Greek alphabet now would be a good time to familiarize yourself with at least some of it.

[13] The following treatment is adequate for our purposes but ignores the (unnecessary here) complexity of covariant and contravariant vectors. However, a good reference is Ref. [12] if one feels the need.

1.5.5   The Non-Cartesian Metric: The Minus Sign

The 4-dimensional space of time and 3-space dimensions is quite different from the Cartesian 3-dimensional space we wander in. One obvious difference is that in space one can go back to where one came from; one cannot go back in time. A related feature of the world is that not all points in 4-space are "reachable" from a given point; the finite velocity of light restricts communication to points for which the value of $|x^\mu|^2 = (ct)^2 - x^2 - y^2 - z^2$ is greater than or equal to zero. For these points there is time enough for light to travel, i.e., one can communicate. For negative values, the distance in space is larger than the light travel time, and so one cannot. These two cases are respectively called "time-like" ($(ct)^2 > |\vec{x}|^2$), i.e., the time difference is larger than the space distance, and "space-like," the spatial distance is the larger. When the two terms are equal, light from the earlier reaches the later; the surface of these points in 4-space is called "the light cone." See Problem 3 of this chapter.

1.6   Lorentz Transformations Between Frames

We close the chapter on Einstein's heuristic Gedanken Experiments with an introduction to Lorentz transformations, in this case transforming the events in Casals' frame to Primrose's frame.[14] In the next chapters we will develop a more elegant and powerful mathematical language to be able to address the transformations of energy and momentum as well as of time and space.

1.6.1   The Transformation Equations From Casals' Frame to Primrose's Frame

Consider an event in one spatial dimension in Casals' frame F, specified by position x and time t. The events as measured by Primrose in frame F' are specified by:

$$\begin{aligned} ct' &= \gamma\,(ct) + \beta\gamma x \\ x' &= \beta\gamma\,(ct) + \gamma x. \end{aligned} \tag{1.9}$$

1.6.2   Conventions for the Units of Time and Space

Units, to be frank, can be painful, and most texts spend far too much time on them. Here we address the units of time and space.

We will predominantly use two systems: 1) the International System (SI), also known as MKS, for meters, kilograms, and seconds. In SI, the unit of time is the

---

[14] We omit derivations of Lorentz transformations beyond the heuristic demonstrations. However, see Problem 4 for a "derivation" of two of the matrix elements and one constraint on the matrix.

second and the unit of space is the meter. Each of these has developed historically and independently; consequently in SI the constant of proportionality c in the 4-vector $(ct, \vec{x})$ has units of meters/seconds.[15] For purely historical reasons the conversion constant between distances in time (in seconds) and space (in meters) has the numerical value 3.0×10^8.[16]

In Natural Units the units of length and time are chosen so that the velocity of light is identically equal to 1, $c \equiv 1$. This may be familiar; astronomers have long chosen the unit of time Δt to be a year, and the unit of distance in space Δl to be a light-year, the distance light travels in one year. The conversion factor c for the speed of light in these units is then unity by definition: $c = \Delta l / \Delta t \equiv 1$.

Working particle physicists exploit an approximate relationship between 1 foot as the unit of distance and 1 nanosecond (10^{-9}s—i.e., one-billionth of a second), giving a value[17] for c within 2% of 1. We consequently will work in nsec and feet for most terrestrial relativistic problems. Note that with $c = 1$ you can measure time in feet or length in nsec; an object 6 feet away is also 6 nsec of light-travel time away, meaning that the light from it had to leave that much earlier to arrive at the same time as light from a nearby object. You are always seeing in the past while awaiting the future.[18]

1.6.3   Putting the Factors of c Back in by Dimensional Analysis

We will use natural units for relativistic problems such as occur naturally in particle physics, cosmology, and astronomy. Note that since $c = 1$, expressions such as v/c and ct become v and t, respectively. "Okay," you say; "it's much cleaner, but how to know where to put the c's back in after you have finished a calculation and want to convert to SI units?" It becomes natural from the context. If you have a t in an expression for length, you need to make it ct. If you have a v in an expression that is dimensionless (i.e., not a length or a time, as occurs in γ), you can make the dimensionless velocity β by dividing by c: $\beta \equiv v/c$.

As an example, the transformation of an event in Casals' frame F to Primrose's frame F' of Eq. 1.5 is given in natural units by:

$$
\begin{aligned}
t' &= \gamma t + \beta \gamma x \\
x' &= \beta \gamma t + \gamma x.
\end{aligned}
\tag{1.10}
$$

In summary, in natural units $c \equiv 1$ and we will not write it explicitly when it is a multiplier or divisor. If you want to convert back to SI, wherever there is a t as a distance multiply by $c = 3 \times 10^8$ m/s to convert seconds to meters, and likewise wherever there is a β multiply by c to get v in m/s. Not hard.

---

[15] So that when the time in seconds is multiplied by c one gets the number in meters.

[16] We work to 2 significant figures, one more than is actually needed here.

[17] One foot is 30.48 cm; light travels 29.98 cm in 1 nsec; the ratio is 1.017. Working to two significant figures, we take c as 1.0.

[18] Which may already have left its source and be on its way.

1.7   Problem Set 1: Vectors, Time Dilation, Lorentz Contraction, Simultaneity, and the Lorentz Transformation

Time Management and Study Groups: You need to work with your study group. The problem sets will go faster if you discuss the problems, with friends/colleagues, and you will have a deeper understanding. However, the work you hand in **has to be your own.**[19]

Problems with answers, and recycled problems: There are a limited number of easily-solved mechanics problems, and so one can find answers to most by searching on the web. We trust you to instead work them yourself; ask your fellow students, the TA's, and/or your instructor for help if you need it. Browsing other texts is recommended; however, you should write out the solution with the book closed.

Getting help: Yes, if you need help ask for it. Bring your study group with you. Do it more than once if needed.

Formulae: For the velocity β and the Lorentz factor γ:

$$\beta = v/c; \quad \gamma^2 = 1/(1 - \beta^2); \quad \beta^2 = (\gamma^2 - 1)/\gamma^2; \quad For \ \gamma >> 1, \ \beta \approx 1 - \frac{1}{2\gamma^2}.$$

The invariant length of the 4-vector x_0, x_1, x_2, x_3: $|x^\mu| = \sqrt{x_0^2 - x_1^2 - x_2^2 - x_3^2}$. Lorentz transformation for a "Boost" of frame F along the x direction relative to frame F':

$$t' = \gamma t + \beta \gamma x \qquad (1.11)$$

$$x' = \beta \gamma t + \gamma x \qquad (1.12)$$

$$y' = y \qquad (1.13)$$

$$z' = z \qquad (1.14)$$

$$\begin{pmatrix} t' \\ x' \\ y' \\ z' \end{pmatrix} = \begin{pmatrix} \gamma & \beta\gamma & 0 & 0 \\ \beta\gamma & \gamma & 0 & 0 \\ 0 & 0 & 1 & 0 \\ 0 & 0 & 0 & 1 \end{pmatrix} \begin{pmatrix} t \\ x \\ y \\ z \end{pmatrix}. \qquad (1.15)$$

Problems: Solutions will be provided.[20] *Please do not plug in any numerical values until the end.* In Problems 1 and 2 not all parts need be assigned if the set is deemed too long.

---

[19] I once required two students who inadvertently strayed to read Egil Krogh's book *Integrity* (see Skills and Guidelines).

[20] Having high quality solutions available at the problem set submission deadline is essential feedback. Do not settle for less.

Problem 1: Practice with 3-Vectors[21]

Consider the two vectors $\vec{A} = (-3, 1, -2)$ and $\vec{B} = (2, -2, 3)$ respectively:

1. Calculate the length of $\vec{A}$; (don't bother explicitly taking the square root, it's quicker to leave the length squared under the sqrt sign);

2. Calculate the length of $\vec{A} + \vec{B}$;

3. Calculate the length of $\vec{A} - \vec{B}$;

4. Draw a diagram of the reference frame showing the x, y, and z axes and the position vectors $\vec{A}$ and $\vec{B}$;

5. On your diagram show $\vec{A} - \vec{B}$ and $\vec{A} + \vec{B}$;

6. Calculate $\vec{A} \cdot \vec{B}$;

7. Calculate the angle between $\vec{A}$ and $\vec{B}$;

8. Calculate the projection of $\vec{A}$ on $\vec{B}$;

9. Calculate $\vec{A} \times \vec{B}$;

10. Find $(\vec{A} \times \vec{B}) \cdot \vec{A}$;

11. Find $(\vec{A} \times \vec{B}) \times (\vec{A} \times \vec{B})$.

Problem 2: Indices and Conventions

1. Define "Index" in the context of vectors and matrices and write down examples with 0, 1, 2, 3, and 4 indices, respectively (not trivial—discuss with your group).

2. Prepare a 2-minute semi-formal talk for your study group on what an index is and isn't. If you use Powerpoint or equivalent it should be no more than 1 slide.

3. Show that

$$\vec{A} \cdot \vec{B} = \sum_{i=1,3} A_i B_i. \tag{1.16}$$

4. Show that

$$\vec{A} \cdot \vec{B} = \sum_{i=1,3} \sum_{j=1,3} A_i B_j \delta_{ij}, \tag{1.17}$$

where δ_{ij} is the "Kronecker delta" (see Section A.1.8 of Appendix A.) If you are bothered or confused by the problem, write out all 9 terms.

5. Show that

$$(\vec{A} \times \vec{B})_i = A_j B_k - A_k B_j \tag{1.18}$$

---

[21] See Section A.1.1 of Appendix A for the scalar product. Also, if you are bold, Appendices A.1.8 and A.1.9 for the vector product and an elegant notation for both.

and cyclic ($i \rightarrow j \rightarrow k \rightarrow i$).

6. Show that

$$(\vec{A} \times \vec{B})_i = \epsilon_{ijk} A_j B_k \tag{1.19}$$

and cyclic ($i \rightarrow j \rightarrow k \rightarrow i$) where ϵ_{ijk} is the Levi-Civita tensor (see Section A.1.8 of Appendix A). If you are bothered or confused by the problem, write out all 27 terms.

Problem 3: 4-Vectors and the Invariant Length

1. Consider the 4-vectors $x^\mu = (t, x, y, z) = (13, 0, 12, 5)$, $(0, 3, -6, -5)$, and $(-6, 0, -3, -2)$ where time is measured in nsec and space coordinates in feet. What is the invariant length squared of each?

2. Consider two events at space-time points $A^\mu = (15, 4, -16, 7)$ and $B^\mu = (2, 4, -4, 2)$ respectively, where time is measured in nsec and space coordinates in feet. What is the invariant distance squared in space-time between them, $|B^\mu - A^\mu|^2$? What is the distance in space between them? In time?

3. Suppose event A happened at time $t_A = 16$ nsec rather than 15. What is the distance in space between A and B? In time?

4. Suppose event A happened at time $t_A = 14$ nsec rather than 15. What is the distance in space between A and B? In time?

5. In each of the above three examples can event A cause event B?

Problem 4: The Lorentz Transformation of an Event in Space-Time

The Lorentz transformation for a boost of an event in frame F along the x direction relative to frame F' is given by[22]

$$t' = \gamma t + \beta \gamma x \tag{1.20}$$

$$x' = \beta \gamma t + \gamma x \tag{1.21}$$

where γ is the Lorentz factor $\gamma = \frac{1}{\sqrt{1 - \beta^2}}$, and β is the velocity in Natural Units, $\beta = v/c$.

1. Find the time t' in frame F' for an event at the location $(t, 0)$ (i.e., at time t at the origin) in frame F. Which of the Einstein Gedanken experiments does this

---

[22] We (naturally) work in natural units (NU). Distances in the first coordinate are measured in nanoseconds (10^{-9} seconds); distances in the next 3 coordinates are measured in feet. The speed of light (good enough for government work) is 1 ft/nsec, i.e., $c = 1$. If this troubles you, put a "c" next to every "t" where $c = 3 \times 10^8 \ m/sec$, and work in SI units. You will get over it.

correspond to, and to which special point in frame F does $x = 0$ correspond? (A trivial question, but meaningful.)

2. Find the location x' in frame F' for an event at the location $(t, 0)$ (i.e., at time t at the origin) in frame F. Please parse this in terms of distance = velocity times time.

3. The structure of the transformation matrix for the Lorentz transformation, represented by Eq. 1.20 with γ as the diagonal elements and $\beta\gamma$ as the off-diagonal elements, is not easy to wrap one's mind around. Show that the invariant length of the event 4-vector position is the same in frames F' and F, i.e.,

$$|x'^{\mu}|^2 = |x^{\mu}|^2$$
$$t'^2 - x'^2 = t^2 - x^2. \tag{1.22}$$

Problem 5: Cosmic Rays[23]

Consider a muon (a heavy cousin of the electron), identical in the form of its interactions with matter except for effects due to its being 200 times heavier,[24] created in the atmosphere by a cosmic ray coming from far away. Assume that in its own rest frame, this individual muon has a lifetime of $\tau = 2200$ nanoseconds,[25] after which it decays to a muon neutrino and an electron/anti-neutrino pair. This muon is traveling with velocity $\beta = v/c = 0.9999995$ ($\gamma = 1000$) with respect to the Earth.

1. Draw a clear (not too small) diagram of the process in the muon rest frame and another diagram in your own frame. Be sure to label the respective origins and axes.

2. Write down the 4-vector for the decay point in the coordinate frame of the muon.

3. Starting with the value of β, calculate the Lorentz factor γ for the transformation from the muon frame to the Earth frame.

4. Lorentz transform the 4-vector representing the decay point in the muon frame to get the 4-vector for the decay point in the Earth's frame.

5. How long is the lifetime as measured in the Earth's frame?

6. How far did the muon travel from where it was created to where it decayed in the Earth's frame?

7. How far would the muon have traveled without the factor of γ?

8. Calculate the proper time (the invariant length of the 4-vector) from the coordinates of the decay event in both the muon and Earth's frame.

---

[23] On a personal note, I recommend the (oldie) film: Time Dilation an Experiment with Mu Mesons 1962 PSSC David Frisch, James Smith, MIT. I found it on YouTube.

[24] I. I. Rabi (Columbia Univ.) is famously quoted as saying, "Who ordered *that?!*"

[25] The distribution in how long muons live goes as $e^{-\frac{t}{\tau}}$, where $\tau = 2200$ nanoseconds.

Problem 6: Time Dilation

Consider the first Einstein Gedanken Experiment. A simple clock is constructed on a *very* fast Chicago Metra Electric train by mounting an LED (light-emitting diode) and a photo-diode together inside the train on one wall, and a mirror on the wall across the train and directly opposite. The LED and photodiode are pointed at the mirror, and are electrically connected so that a short LED pulse reflected from the mirror triggers the photodiode to make the LED flash. The result is that the LED flashes repeatedly at a fixed interval that corresponds to twice the light transit time across the width of the train. The train is an Express, moving at $\beta = 0.99995$, ($\gamma = 100$) relative to the station.

1. Set up the problem and define the relevant events in the frame of the train. (Be sure to draw a well-labeled clear diagram.)

2. Transform the coordinates of each event into the frame of the station.

3. Draw a carefully-labeled diagram of the geometry of the light path in the frame of the station.

4. Find the time between flashes as seen in the frame of the station.

5. Find the distance between flashes as seen in the frame of the station.

Problem 7: Einstein Gedanken Experiment 3: The Frame-dependence of Simultaneity

Casals is on a train moving at speed corresponding to a Lorentz factor of $\gamma = 1000$ down a set of tracks past a platform on which Primrose is standing. Casals is in the middle of the train, i.e., equidistant from both ends. Just as Casals is opposite Primrose[26] each of them sees two simultaneous flashes of light that were produced by a light at each end of the train. Casals measures the length of the train to be L. Ignore the width of the train as the length is much longer than the width.

1. Draw a picture and label the frames and axes.

2. Taking the origins of the two coordinate systems and clocks to be the point where Primrose and Casals are when they see the flash, write down the 4-vector in Casals' reference frame corresponding to the position of each light when it flashed.

3. Use the Lorentz transformation to find the 4-vectors of each light when it flashed in Primrose's frame.

4. Primrose can calculate the length of the train from the following reasoning: the spatial separation of the two flashes is the distance the back of the train moved

---

[26] Take them to be so close as to effectively be at the same location.

while the light was propagating plus the length of the train. In symbols,

$$\Delta x' = \beta \Delta t' + L'. \tag{1.23}$$

Find the length of the train as measured by Primrose, L' in terms of L and γ. (Remember (learn) the identity $1/\gamma^2 = (1 - \beta^2)$).

5. Casals deduces that the two lights flashed simultaneously. In contrast, Primrose claims they had to have flashed at different times for him to have seen the flashes simultaneously. What is the time interval between the two lights flashing in Primrose's frame?" (The perils of translating into English—a better way to have asked is "Transform the two light-flashing events into Primrose's frame and find the time difference.")

CHAPTER 2

Transformations as Operators Acting on Vectors in a Space

2.1   Introduction

In Chapter 1 we used the three Einstein Gedanken experiments to develop some intuition that takes into account that the speed of light is not infinite. However, although the Einstein examples are wonderful pedagogy, they are carefully crafted and are not generalizable. In this chapter we will develop a mathematical language that allows transformation between frames of reference in any direction, and that supports successive transformations in different directions and different velocities.[1]

2.2   Events as Points in Time and Space

Consider some action that happens at a well-defined point in time and space, for example, the flash of the light flasher, the bounce at the mirror, or the detection of a flash by Casals or Primrose. We will define each of these as an *event*. A time-ordered list of events is a familiar concept in daily life. However, as we saw heuristically in the Gedanken Experiments, a list of events, when and where, in one frame can be quite different in another. In solving the problems on the Problem Sets please be systematic in writing down the sequence of events in one frame before trans-forming into another. Instead of trying to hold a mental picture of multiple events occurring at different times in separated locations, a picture with which we have no experience, making a list of events in one frame and then transforming each event into another frame is "sure-footed". One doesn't have to think beyond one step-at-a-time. From the new list we can then easily calculate the time intervals and distances between events in the transformed frame.

---

[1] The language uses only multiplication and addition; no calculus is needed. However, it does use arrays of numbers arranged as vectors and matrices.

Let us begin on the mathematics of the transformation of a sequence of events from one frame to another (the recipe). First, create the list of events in one frame. The steps are listed below.

1. Define a frame of reference for the problem as stated, with an origin, axes, and units in a 4-dimensional space (time and 3-space).

2. Identify the events in the problem.

3. Define each event by its time t and place $\vec{x}$.

4. We will use the convention that a point in this space is described by a 4-vector $x^\mu \equiv (ct, x, y, z)$, where the super-script μ (mu) is the Greek letter for m, and can take on the values of 0,1,2,3. We call the label of the component the index. Here we have used the letter μ, but in our convention for 4-space we can use any Greek letter as the index and have the meaning understood.

5. The frame of reference in which the object is not moving is called its rest frame. If the problem involves a particle (an object with small spatial extent, such as an electron or even a small ball), we typically use the convention that in the rest frame the particle is always located at the origin, and so only the time component is not zero.

6. In 3-space, we use the convention that for indices we use Roman letters as subscripts that run from 1 to 3. The 3-vector for a position in space is $\vec{x} = (x_1, x_2, x_3)$.

Each event is thus fully characterized by the 4 coordinates (ct, x, y, z) in a chosen frame. Given the four coordinates of two events in the same frame one can calculate the time or length between them.[2]

As we saw in the three Gedanken experiments of Chapter 1, Casals and Primrose both have self-consistent histories[3] of what happened, but disagree on when and where, i.e., the 4 coordinates of each event. Casals in frame F characterizes a certain event by (ct, x, y, z), and Primrose in F' characterizes the same event by (ct', x', y', z'), where in general the values of each component are different in the two frames.

2.3   Lorentz Transformations, Indices, Vectors and Matrices

Here we develop a mathematical language for transformations of events between frames that is concise, easier to apply, and in which we will see the mathematics taking on a life of its own. We are limiting ourselves to inertial frames, i.e.,

---

[2] Note that by using ct for the zeroth component all four coordinates have units of length.

[3] Time-ordered lists of events.

frames moving relative to each other at uniform velocity, with no accelerations. The transformations between coordinate systems are then linear as described below.

2.3.1   Transforming Events from One Frame to Another

Transforming an event from Casals' frame F to Primrose's frame F' can be described by 4 linear equations in the time and space coordinates of the event. Writing out the equation for each of the components in F' in terms of the 4 components in frame F:

$$ct' = M_{00}\ (ct) + M_{01}\ x + M_{02}\ y + M_{03}\ z$$
$$x' = M_{10}\ (ct) + M_{11}\ x + M_{12}\ y + M_{13}\ z$$
$$y' = M_{20}\ (ct) + M_{21}\ x + M_{22}\ y + M_{23}\ z$$
$$z' = M_{30}\ (ct) + M_{31}\ x + M_{32}\ y + M_{33}\ z.$$

$$(2.1)$$

The 16 coefficients M_{00} to M_{33} are real numbers. The first subscript on each coefficient, for example the 0 in M_{01}, labels the row of the transformation array: all the coefficients in the top row have a first subscript of zero. The second subscript, for example the 1 in M_{01}, labels the columns: all the coefficients in the second column (remember we are counting 0,1,2,3; the second row has a label of 1) have a second subscript of 1.

As a concrete example of a transformation from F to F' with which we are familiar, if we define the motion of the train to be in the x direction, events on the train (Casals' frame) as seen from the station platform (Primrose's frame) are described by the 4 equations:

$$ct' =\quad \gamma\ (ct)\ +\beta\gamma x\ +0\,y\ +\ 0\,z$$
$$x' =\quad \beta\gamma\ (ct) +\ \gamma\ x\ +\ 0\,y\ +\ 0\,z$$
$$y' =\quad 0\ (ct)\quad +0\,x\ +\ 1\,y\ +\ 0\,z$$
$$z' =\quad 0\ (ct)\quad +0\,x\ +\ 0\,y\ +\ 1\,z$$

$$(2.2)$$

where β is the velocity in natural units, v/c, and γ is the Lorentz factor. This transformation of the time and space coordinates from one frame into another moving at constant velocity is called a boost.[4] Equation 2.2 is specific to a boost along the x direction, assuming that the origins of the two frames coincide at $t = t' = 0$ with the spatial axes aligned (i.e., x points in the x' direction, etc.).

These four equations may look formidable, but are a succinct way of expressing the recorded behavior of the trains and platforms in the Gedanken Experiments of Chapter 1. The mathematics requires only multiplication and addition.[5]

---

[4] Concise and evocative; I don't know who invented the name (although it is often called a Lorentz boost).
[5] Nothing to fear here.

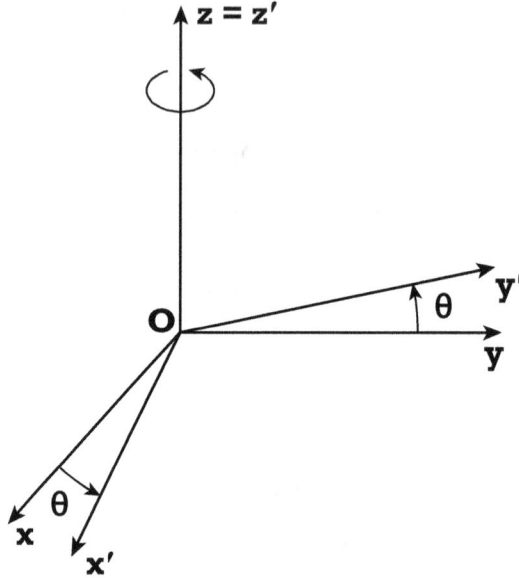

Figure 2.1. Frame F' is rotated about the z-axis with respect to Frame F by an angle θ. Note that we use the convention of rotating the frame rather than the vector; for the other convention, change θ to $-\theta$ and $\sin(\theta)$ to $-\sin(\theta)$.

The top-most line in Equation 2.2 provides the time t' in Primrose's frame F' of an event occurring at time t and position x, y, z in Casals' frame F. The second line provides the position along the x' axis in Primrose's frame of an event occurring at at time t and position x, y, z in Casals' frame F. The third and fourth lines reduce to $y' = y$ and $z' = z$, respectively, and express the invariance of lengths in the dimensions perpendicular to the motion.

If the train were instead running along the platform in the y direction with velocity β, the transformation matrix would "mix" the time and the y coordinate in lines 1 and 3, and leave the x and z components unchanged (lines 2 and 4, respectively):

$$
\begin{aligned}
ct' &= \gamma\ (ct) + 0\ x + \beta\gamma\ y + 0\ z \\
x' &= 0\ (ct) + 1\ x + 0\ y + 0\ z \\
y' &= \beta\gamma\ (ct) + 0\ x + \gamma\ y + 0\ z \\
z' &= 0\ (ct) + 0\ x + 0\ y + 1\ z.
\end{aligned}
\tag{2.3}
$$

2.3.2   Rotations as a 3 × 3 Subset of the Transformation in 4-Dimensions

The same framework of 4 equations describes spatial rotations. As an example, consider Casals to be in frame F as before, described by ct, x, y, and z, and Primrose to be in frame F', rotated about the z-axis as shown in Fig. 2.1. Note that frame F' is not moving with respect to F, and the two origins and z-axes coincide.[6]

$$
\begin{aligned}
ct' &= 1\ (ct) &&+ 0\,x &&+ 0\,y + 0\,z \\
x' &= 0\ (ct) + \cos(\theta)\,x &&+ \sin(\theta)\,y &&+ 0\,z \\
y' &= 0\ (ct) - \sin(\theta)\,x &&+ \cos(\theta)\,y &&+ 0\,z \\
z' &= 0\ (ct) &&+ 0\,x &&+ 0\,y + 1\,z
\end{aligned}
\tag{2.4}
$$

Eliminating the variables which are the same in both frames, we get:

$$
\begin{aligned}
x' &= \cos(\theta)\,x + \sin(\theta)\,y \\
y' &= -\sin(\theta)\,x + \cos(\theta)\,y
\end{aligned}
\tag{2.5}
$$

which describes a rotation of frame F about the z-axis by θ.

2.3.3 Indices

If you are comfortable with indices feel free to skip this section.

We can save time and effort by labeling the four components (ct, x, y, z) by their position in the vector rather than explicitly writing them out. Consider the elements of the first column of Eq. 2.1. Using the convention of counting starting with zero, the left-hand side of each row in turn reads ct', x', y', z'. We thus label the 4 positions by the index $\mu = 0, 1, 2, 3$. The quantity in the zero position is ct', and the values $\mu = 1, 2, 3$ correspond to the three components of $\vec{x}$.

As an example, consider a seat assignment in a large rectangular lecture hall. We can assign an integer i to the row, starting at the front, and an integer j to the seat, starting on the left facing the hall from the front. The identifier for the seat is thus M_{ij}, where M is a 2-dimensional array comprising all the M_{ij}. The indices i and j run from 1 to the number of rows and seats per row, respectively.

Now suppose the lecture hall is in a building of N identical floors, each with a lecture hall identical to the one we've described. We can identify a seat in any one of the halls by M_{ijk}, where i is the row, j is the seat in the row, and the index k is the floor of the building, running from 1 to N.

Now suppose all this is happening on the campus of a large state university with many campuses (UC Berkeley, UCLA, UCSB, UCSC, UC Davis, UC Merced, ...), and that each campus has a similar building. Label the campus by the index l: the seat is now identified by M_{ijkl}.[7] And so forth—an additional index for each of other countries, other planets, other universes....

---

[6] The convention is that events happen in F, and are transformed to F', opposite from many texts. I think of rotations, for which downtown Chicago stays where it is, but I turn to look at it. To use the convention in which the vector is rotated instead of the frame, one has only to change the sign of θ for rotations, and correspondingly, of β for boosts.

[7] Thus M_{3145} is the identifier for row 3, seat 1, floor 4, campus 5, i.e., the first seat in the third row of the lecture hall on the 4th floor of the Physics building at UC Davis. M could be some attribute of the seat such as availability, color, etc.

2.3.4   Vectors

For 4-vectors, such as (ct, x, y, z), the convention is we use lower-case Greek letters for the index values of 0, 1, 2, or 3. For example, if we use μ (mu, the Greek m) as the index, the component of x^μ with index $\mu = 0$ is ct, the component with index $\mu = 1$ is x, and so forth.

The index is a pointer to the position in the vector x^μ; it is an integer, and has a range determined by the size of the array. For now we will also use the convention that for 4-vectors the index is a superscript (upper index), e.g., $x^\mu = (x^0, x^1, x^2, x^3)$ stands for the 4-vector (ct, x, y, z), where, for example, $x^3 = z$.

For 3-vectors, such as $\vec{x} = (x, y, z)$, our convention is we use lower-case Roman letters that stand for values 1 to 3, and we write the index as a subscript. For example, if we use i as the index, the $i = 1$ location in the vector, labeled x_1, is x, the $i = 2$ location x_2 is y, and the $i = 3$ location x_3 is z.[8]

We will represent a vector as either a vertical column or a horizontal row with comma-separated elements, depending on the context.

2.3.5   Matrices

We will use the same index conventions for matrices, with the exception that for now we will use subscripts both in 4-space and 3-space.

In 4-space a matrix is dimensionally 4×4. We label each element by two indices, the row index in the first location followed by the column index, as in $M_{\mu\nu}$, where both μ and ν run from 0 to 3. Note that the 4 elements of a given row, e.g., the $\mu = 0$ row, have the same ordering as a 4-vector, i.e., ct, x, y, z. However, the 16 matrix elements are dimensionless coefficients, not coordinates.

In 3-space, a matrix is (of course) 3×3. We again label each element by two indices, the row index in the first position followed by the column index, as in M_{ij}, where both i and j run from 1 to 3, corresponding in Cartesian coordinates to x, y, and z, respectively.

2.4   Transformations as Operators Acting on Vectors in a Space

We can make an abstraction with remarkable consequences. Let us use the indices of the coefficients $M_{\mu\nu}$ to define the transformation, with the first index being that in Primrose's vector (frame F') and the second that in Casals' vector (frame F). The vectors themselves no longer appear but are implicit in the locations in the

---

[8] Note that with these conventions the values of the index for the spatial coordinates x, y, and z are 1,2,and 3, respectively, the same in 3-space and 4-space. The $\mu = 0$ location in the 4-vector is reserved for the time dimension.

matrices, defined by the indices.[9] The matrices represent operations on the vectors of the space.

The transformation of the events (flashes, bounces) in F to F' in the first Gedanken Experiment is now given by

$$t' = \gamma t + \beta \gamma x \qquad (2.6)$$

$$x' = \beta \gamma t + \gamma x \qquad (2.7)$$

$$y' = y \qquad (2.8)$$

$$z' = z \qquad (2.9)$$

where we have written a boost by velocity β along the x direction as 4 separate equations, 2.6 to 2.9.

We can instead represent the transformation as a matrix of the coefficients acting on a vector in F to produce the corresponding vector in F':

$$\begin{pmatrix} t' \\ x' \\ y' \\ z' \end{pmatrix} = \begin{pmatrix} \gamma & \beta\gamma & 0 & 0 \\ \beta\gamma & \gamma & 0 & 0 \\ 0 & 0 & 1 & 0 \\ 0 & 0 & 0 & 1 \end{pmatrix} \begin{pmatrix} t \\ x \\ y \\ z \end{pmatrix}. \qquad (2.10)$$

As an example, reading across the top row and referring to Section A.1.1.4 in Appendix A, we find for t':

$$t' = \gamma t + \beta \gamma x. \qquad (2.11)$$

2.5   What We Have Gained With a Lorentz-Invariant Language

Our "language" of algebraic Lorentz transformations allows new capabilities, which we list below.

2.5.1   A Comprehensive Description of Boosts and Rotations

We can now do successive boosts in any direction and rotations about any axis, in any order we need. Successive transformations correspond to multiplying the respective matrices together to get the single overall transformation matrix.

---

[9] For example, the M_{00} element is in the first row first column, and the M_{13} element is in the second row fourth column.

2.5.2   The Transformation Properties of 4-Vectors

All 4-vectors transform from one frame to a moving and/or rotated frame according to Equation 2.1. In this course we will be working with the 4-vectors $x^\mu = (ct, x, y, z)$ and, soon, $p^\mu = (E, cp_x, cp_y, cp_z)$ to solve a wide variety of kinematic problems. Using 4-vectors and Lorentz transformations we can choose the most appropriate frame, easily perform multiple successive transformations, and exploit invariants between frames, working in a powerful symbolic language.

In addition, although not our topic here, the electromagnetic charge density and current form a 4-vector, $J^\mu = (c\rho, J_x, J_y, J_z)$, and the electromagnetic scalar and vector potentials, $A^\mu = (V/c, A_x, A_y, A_z)$, from which the electromagnetic fields $\vec{E}$ and $\vec{B}$ are derived, form another. The 4-vectors for the potentials and the currents transform by the identical prescription for the 4-vectors for position and momentum.[10]

2.5.3   The Matrices Represent a Group

Consider 4-space. The set of 4×4 matrices of boosts and rotations form a mathematical group, satisfying the three conditions of a group:

1. The product of two Lorentz transform matrices is also a Lorentz transform;

2. There exists an Identity matrix;

3. For each matrix there is an inverse matrix.

The Identity matrix consists of 1's on the diagonal and zeros off-diagonal. The inverse of the matrix for a boost is formed by changing the sign of the relative velocity, $\beta \to -\beta$.

The 3×3 matrices formed by the space components also form a group, in this case purely rotations. The inverse is formed by changing the sign of the angle of rotation, $\theta \to -\theta$.

2.5.4   Parts of Speech: Lorentz Scalars, Vectors, and Tensors

We have taken as our starting point the principle that the description of physical phenomena must be the same in all inertial frames. This is a strong constraint on the descriptive language: the equations governing motion and electrodynamics, for example, need to be expressed in a language that remains invariant under boosts and rotations between frames.

---

[10] The vector potential $\vec{A}$ is often given short shrift in first-year physics courses, appearing at the end of a quarter as something arcane and difficult. It's neither, but is instead the space components of the 4-vector for electromagnetic potential, just as $\vec{x}$ are the space components associated with time. Once one has the potentials, the electric and magnetic fields are obtained by taking derivatives. This is a powerful example of Mark Twain's description of science as "such wholesale returns of conjecture out of such a trifling investment of fact." [11]

Table 2.1. The "Parts of Speech" of our Lorentz invariant language. As an example of the power of a systematic principle-based approach, once one has mastered the transformation of the 4-vector for time and space under boosts and rotations, one has also mastered the transformations of the 4-vector for energy and momentum, the 4-vector for the electric charge and current, and the 4-vector for the scalar and vector electromagnetic potentials. The electric and magnetic fields are easily derived from the potentials.

Type	Examples	Comment
Lorentz Scalar	$3, \pi, \vec{A} \cdot \vec{B}, |x^\mu|^2$	Numbers
Lorentz Vector	$x^\mu, p^\mu, A^\mu, J^\mu$	4-Vectors
Lorentz Transform	$M_{\mu\nu}$	Boosts and Rotations

The parts of speech we will use in this introductory course fall into 3 categories,[11] summarized in Table 2.1:

2.6   Successive Lorentz Transformations; an Example

A boost of $\beta = 0.9c$ in the x direction followed by another boost also of $0.9c$ if added directly would seemingly violate the velocity limit of $c = 1$. To illustrate the representation of successive Lorentz transformations by the product of their respective matrices, we will use the product of two successive boosts to derive the correct behavior for adding velocities.

2.6.1   The Law of Addition of Velocities

Consider the case of two successive boosts in the same direction, as shown in Figure 2.2. Frame F is moving at velocity β_1 in frame F', which itself is moving at velocity β_2 in frame F''. We want to solve for the velocity of frame F in frame F''; call it β_T (T for Total). All motions are along the respective x-axes.

The successive Lorentz transformations are represented by:

$$\begin{pmatrix} t'' \\ x'' \\ y'' \\ z'' \end{pmatrix} = \begin{pmatrix} \gamma_2 & \beta_2\gamma_2 & 0 & 0 \\ \beta_2\gamma_2 & \gamma_2 & 0 & 0 \\ 0 & 0 & 1 & 0 \\ 0 & 0 & 0 & 1 \end{pmatrix} \begin{pmatrix} t' \\ x' \\ y' \\ z' \end{pmatrix} \qquad \begin{pmatrix} t' \\ x' \\ y' \\ z' \end{pmatrix} = \begin{pmatrix} \gamma_1 & \beta_1\gamma_1 & 0 & 0 \\ \beta_1\gamma_1 & \gamma_1 & 0 & 0 \\ 0 & 0 & 1 & 0 \\ 0 & 0 & 0 & 1 \end{pmatrix} \begin{pmatrix} t \\ x \\ y \\ z \end{pmatrix}.$$

$$\text{(2.12)} \qquad\qquad\qquad\qquad\qquad \text{(2.13)}$$

Abstracting the transformations from the underlying vector spaces, the combined transformation is given by the product of the two transformation matrices:[12]

---

[11] The Kronecker and Levi-Civita tensors are defined in Section A.1.1 of Appendix A.
[12] Note the order—the matrices operate to their right.

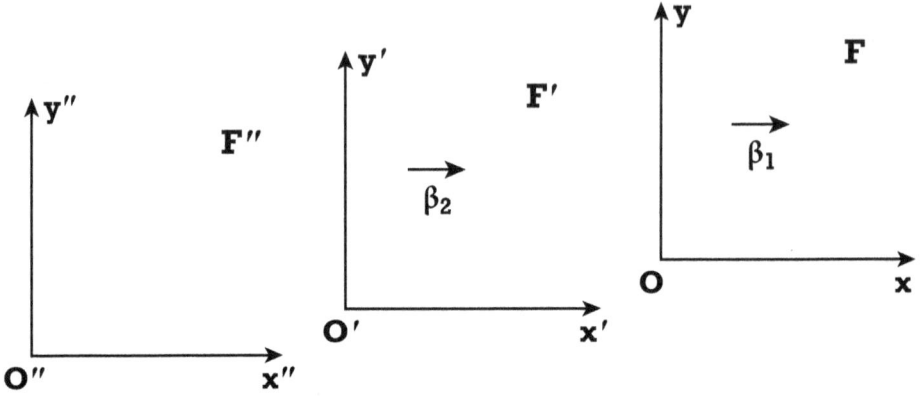

Figure 2.2. Two successive boosts in the same direction. Frame F is moving at velocity β_1 in frame F', which itself is moving at velocity β_2 in frame F''. All motions are along the respective x-axes.

$$\begin{pmatrix} \gamma_T & \beta_T\gamma_T & 0 & 0 \\ \beta_T\gamma_T & \gamma_T & 0 & 0 \\ 0 & 0 & 1 & 0 \\ 0 & 0 & 0 & 1 \end{pmatrix} = \begin{pmatrix} \gamma_2 & \beta_2\gamma_2 & 0 & 0 \\ \beta_2\gamma_2 & \gamma_2 & 0 & 0 \\ 0 & 0 & 1 & 0 \\ 0 & 0 & 0 & 1 \end{pmatrix} \begin{pmatrix} \gamma_1 & \beta_1\gamma_1 & 0 & 0 \\ \beta_1\gamma_1 & \gamma_1 & 0 & 0 \\ 0 & 0 & 1 & 0 \\ 0 & 0 & 0 & 1 \end{pmatrix}.$$

$$(2.14)$$

Multiplying the two transformation matrices (see Appendix A, Section A.1.1) produces

$$\begin{pmatrix} \gamma_T & \beta_T\gamma_T & 0 & 0 \\ \beta_T\gamma_T & \gamma_T & 0 & 0 \\ 0 & 0 & 1 & 0 \\ 0 & 0 & 0 & 1 \end{pmatrix} = \begin{pmatrix} \gamma_1\gamma_2 + \beta_1\beta_2\gamma_1\gamma_2 & \gamma_1\gamma_2\beta_1 + \gamma_1\gamma_2\beta_2 & 0 & 0 \\ \gamma_1\gamma_2\beta_2 + \gamma_1\gamma_2\beta_1 & \gamma_1\gamma_2 + \beta_1\beta_2\gamma_1\gamma_2 & 0 & 0 \\ 0 & 0 & 1 & 0 \\ 0 & 0 & 0 & 1 \end{pmatrix} \quad (2.15)$$

where γ_T and β_T are respectively the Lorentz factor and velocity for the total transformation.

Reading off γ_T and β_T from the (0,0) and (0,1) locations of the matrix for the total:

$$\gamma_T = \gamma_1\gamma_2 + \beta_1\beta_2\gamma_1\gamma_2$$
$$= \gamma_1\gamma_2(1 + \beta_1\beta_2) \qquad (2.16)$$

$$\beta_T\gamma_T = \gamma_1\gamma_2\beta_2 + \gamma_1\gamma_2\beta_1$$
$$= \gamma_1\gamma_2(\beta_1 + \beta_2). \qquad (2.17)$$

Solving for β_T:

$$\beta_T = \frac{\gamma_T\beta_T}{\gamma_T}$$

$$= \frac{\gamma_1 \gamma_2 (\beta_1 + \beta_2)}{\gamma_1 \gamma_2 (1 + \beta_1 \beta_2)} \tag{2.18}$$

$$\beta_T = \frac{\beta_1 + \beta_2}{1 + \beta_1 \beta_2}.$$

Summarizing, the sum β_T of velocities β_1 and β_2 in the same direction is:

$$\beta_T = \frac{\beta_1 + \beta_2}{1 + \beta_1 \beta_2}. \tag{2.19}$$

2.6.1.1   *Comments on the Matrix Algebra*

This may seem very complicated. However, it is just counting, with the only operations being simple multiplication and addition. The multiplication itself is by rote: the $M_{\mu\nu}$ component of the product is the scalar product of the μth row of the second (left-most) matrix and the νth column of the first (right-most), both treated as simple Cartesian vectors.

Note that all the action in this example (Eq. 2.14) is in the 4 components of the 2×2 matrix that mixes ct and x in the top left-hand corner of the Lorentz transformation; there is no need to even write down the other rows and columns as they form a 2×2 Identity matrix.

2.6.1.2   *The Addition Does Not Allow Super-luminal Velocities*

By inspection, the answer has the correct limiting behavior for large velocities. For example, take the case in Section 2.6 of $\beta_1 = 0.9$ and $\beta_2 = 0.9$.

$$\beta_T = (0.9 + 0.9)/(1 + (0.9)(0.9))$$
$$= 1.8/1.81 \tag{2.20}$$
$$\approx 0.99.$$

Fast, but not greater than 1.

2.7   The Invariant Length of a Vector

In 3-space, the length of a vector with components $\vec{x} = x, y, z$ is $|x| = \sqrt{(x^2 + y^2 + z^2)}$. We will use the convention that we compute the length squared, $|x|^2$, to spare ourselves square root signs, but will often refer colloquially to the length when computing length squared. The length of a 3-vector is invariant under rotations.

In 4-space, it's more complicated as we are keeping track of time as well as position. The invariant length of the 4-vector $x^\mu = (ct, x, y, z)$ is defined[13] by:

$$|x^\mu|^2 = (ct)^2 - x^2 - y^2 - z^2. \tag{2.21}$$

In natural units (NU), in which we measure lengths in either feet or in nsec (nanoseconds):

$$|x^\mu|^2 = t^2 - x^2 - y^2 - z^2. \tag{2.22}$$

The length of a 4-vector is invariant under boosts and rotations, i.e., all Lorentz transformations.

2.7.1   Time-like and Space-like Intervals Between Two Events

Equation 2.22 has the property that the invariant length-squared $|x^\mu|^2$ can be either positive or negative. Two events separated by a positive value of $|x^\mu|^2$ have a time separation greater than the space separation. A signal can travel from the earlier to the later event and so is able to cause it. For example, consider two successive events, $x^\mu = (t_1, x_1, 0, 0)$ and $x^\nu = (t_2, x_2, 0, 0)$. A signal propagating at $v = (x_2 - x_1)/(t_2 - t_1)$, where $v < c$, i.e., $v/c = (x_2 - x_1)/c(t_2 - t_1) < 1$, can travel from the first to the second. We call intervals with $|x^\mu|^2 > 0$, *time-like*.[14]

In contrast, two events separated by a negative value of $|x^\mu|^2$ have a space separation greater than time separation. A signal traveling at the speed of light or less cannot travel from one to the other. For example, again consider two successive events $x^1 = (t_1, x_1, 0, 0)$ and $x^2 = (t_2, x_2, 0, 0)$, but with the space separation $x_2 - x_1$ being larger than the time separation $t_2 - t_1$ (again in units with $c = 1$). Signals cannot travel between them without exceeding c; the two events cannot be causally related. We call intervals with $|x^\mu|^2 < 0$, *space-like*. Note that both space-like and time-like intervals are expressed as positive numbers, for example the separation in Natural Units will be in feet or nsec. The minus sign keeps track of whether the interval between events is time-like or space-like, and in this convention does not introduce imaginary numbers in the interval between two events.[15]

The third case, $|x^\mu|^2 = 0$, corresponds to $\Delta t^2 = \Delta \vec{x}^2$, for example, two events 1 nsec apart in time and 1 foot apart in space. These can be connected only by a signal traveling at the speed of light. The surface of such points in 3-dimension space is called the *light cone*.[16]

---

[13] There are many conventions for the notation to support the relative minus sign between the space and time components squared. Here we use that of Ref. [12]. For our purposes a notation using upper or lower indices does not matter as long as we take the invariant length squared as prescribed here.

[14] A mnemonic is that a time-like interval has the time-difference *larger* than the space-distance (in Natural Units).

[15] Quoting the square, $|x^\mu|^2$, rather than $|x^\mu|$, avoids the issue entirely, and the minus sign makes clear whether or not the relationship between the two events can be causal.

[16] At this point it is *de rigueur* to show a figure of the light cone in 1 time and 1 space dimension. I find that, with the exception of the Twin Paradox (Problem 9 of Section 2.8) which involves more than two frames, it adds nothing to the algebra. You can look it up.

2.8   Problem Set 2: Transforming Events, Histories in Different Frames; Rotations and Boosts

Please start early on these with your study group, and please don't be apprehensive. By the end of the quarter you will be comfortable with SR and the math, and will do well in the course. Work together! The course is based on collaborative learning—teaching each other, asking questions, discussing. There are many unanswered questions underlying our current very-limited understanding of our Universe.

Formulae: Formulae for the velocity β and the Lorentz factor γ:

$$\beta = v/c; \quad \gamma^2 = 1/(1-\beta^2); \quad \beta^2 = (\gamma^2-1)/\gamma^2; \quad \text{For } \gamma \gg 1, \beta \approx 1 - \frac{1}{2\gamma^2}.$$

The invariant length (squared) of the 4-vector $|x^\mu| = x_0, x_1, x_2, x_3 \equiv (t, \vec{x})$:

$$|x^\mu|^2 = t^2 - |\vec{x}|^2. \tag{2.23}$$

Lorentz transformation for a boost of frame F along the x direction relative to frame F' written as 4 separate equations, one for each component of the 4-vector:

$$t' = \gamma\, t + \beta\gamma\, x \tag{2.24}$$

$$x' = \beta\gamma\, t + \gamma\, x \tag{2.25}$$

$$y' = y \tag{2.26}$$

$$z' = z. \tag{2.27}$$

The same 4 equations written in matrix notation:

$$\begin{pmatrix} t' \\ x' \\ y' \\ z' \end{pmatrix} = \begin{pmatrix} \gamma & \beta\gamma & 0 & 0 \\ \beta\gamma & \gamma & 0 & 0 \\ 0 & 0 & 1 & 0 \\ 0 & 0 & 0 & 1 \end{pmatrix} \begin{pmatrix} t \\ x \\ y \\ z \end{pmatrix}. \tag{2.28}$$

Problem 1: Using the Lorentz Transformation—The First Einstein Gedanken Experiment

Consider the first Einstein Gedanken Experiment we discussed in Section 1.4.1. Transform the three events (Figure 1.3) that are the basis of Casals' clock into Primrose's frame.

1. Draw a clear neat diagram of the problem;

2. List the events in Casals' frame;

Event 1
Front of pole at
front door

$(t, x) = (0, 0)$

Event 2
Front of pole at
back door

$(t, x) = (w/v, 0)$

Event 3
Back of pole at
front door

$(t, x) = (L/v, -L)$

Event 4
Back of pole at
back door

$(t, x) = ((L + w)/v, -L)$

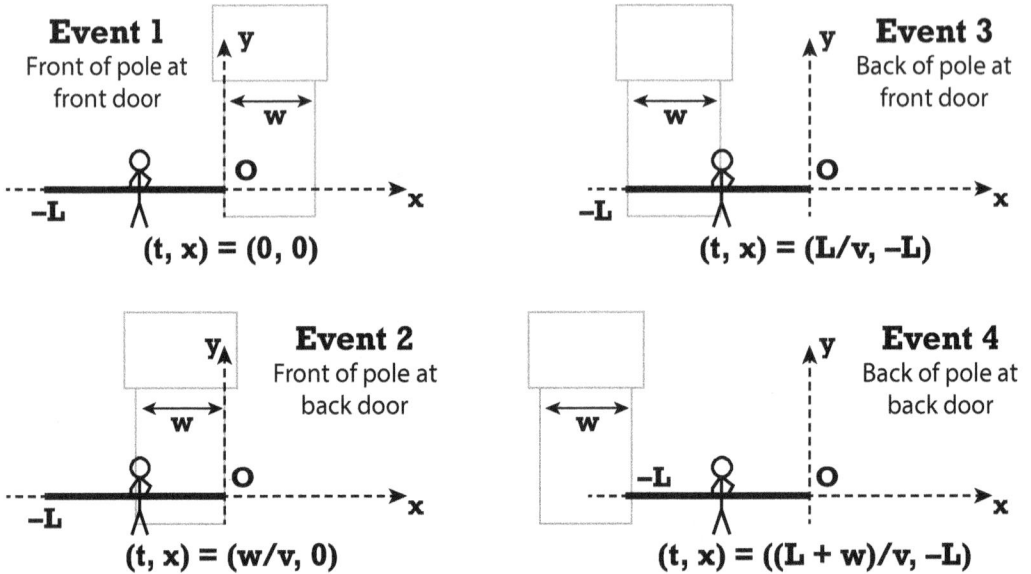

Figure 2.3. The consecutive events in Problem 2 as recorded by the runner. The barn's width to be crossed is w (unprimed) in the runner's frame. We are given that the width of the barn in the farmer's frame, w′, is 30 feet. The pole has a length L= 15 feet in the runner's frame and length L′ in the farmer's frame.

3. For each event, write the 4-vector;

4. Transform the 4-vectors into Primrose's frame;

5. Calculate the distance between the first flash and the second flash in both Casals' and Primrose's frame.

6. Calculate the time between the first flash and the second flash in both Casals' and Primrose's frame.

Problem 2: Using the Lorentz Transformation—The "Barn and the Pole Paradox"

Consider the following folk fable: A runner carrying a long pole horizontally runs through a barn, entering one end and exiting out the other end. A farmer is standing in the barn. The pole is 15 feet long as measured in the runner's frame. The barn is measured to be 30 feet from one end to the other by the farmer.

The barn has a door on each end, controlled by light sensors so that each door opens when the front of the pole reaches it and shuts just after the back end of the pole has gone by.

This is a very fast runner, moving at $\gamma = 10$, $\beta = 0.995$. A paradox is that the farmer sees the runner and the pole as Lorentz-contracted so that at least for a time the pole was inside the barn with both doors closed. However, the runner sees the barn as Lorentz-contracted to be much shorter than the pole, and so there's no way both doors could be closed at the same time with the pole inside. Show that this apparent paradox is entirely natural[17] by calculating the arrival of the front and back ends of the pole at each of the doors in each frame.

Step-wise:

1. Draw a picture using our convention that the primed frame is the frame of the barn, showing the two coordinate frames, with origins and labeled axes. Label the lengths of the barn and pole appropriately. I suggest taking the origins of both frames to be $t = t' = 0$ and $x = x' = 0$ when the front of the pole arrives at the first door.

2. Write down the 4-vectors for the following four events in the runner's frame:

   a. Front end of pole reaches the barn front door;

   b. Front end of pole reaches the barn back door;

   c. Back end of pole reaches the barn front door;

   d. Back end of pole reaches the barn back door.

3. Using the transformation for a boost along the x-direction, transform the 4-vectors for the four events into the frame of the barn.

4. Using appropriate numerical values, order the four events in time in the runner's frame.

5. Using appropriate numerical values, order the four events in time in the frame of the barn.

6. In each of the 2 frames make a short summary of the time sequence of the 4 events. There will be a prize[18] for inventing and performing a test to prove which history is correct.

Problem 3: Indices, Unit Vectors, and Projection Operators

1. In Cartesian 3-space, give the explicit representation of the unit vectors $\hat{x}, \hat{y}, \hat{z}$.

2. Show that $\hat{x}_i \cdot \hat{x}_j = \delta_{ij}$, where the Kronecker δ_{ij} is equal to 1 if $i = j$ and is 0 otherwise (note the indices i and j run from 1 to 3).

3. Show that $\sum_{i=1}^{3} \sum_{j=1}^{3} \delta_{ij} = 3$ (this is the sum of the diagonal elements, called the trace, of the matrix δ_{ij}. See Appendix A, Section A.1.1).

4. Find the unit vector in the $\vec{A}$ direction for $\vec{A} = (2, 3, -5)$.

5. Using the geometric relationship $\vec{A} \cdot \vec{B} = |A||B|cos(\theta)$, show that the vector operator $\hat{n} \cdot$ (n-hat dot) applied to a vector $\vec{A}$ returns the projection of $\vec{A}$ along the $\hat{n}$ direction.

---

[17] Natural!
[18] Nobel.

Problem 4: Cross and Scalar Products in Tensor Notation

A challenge: the Levi-Civita tensor (see Section A.1.8 of Appendix A). Too advanced at this moment in the course. However, it's nice.

Prove the tensor identity:

$$\epsilon_{ijk}\epsilon_{ilm} = \delta_{jl}\delta_{km} - \delta_{jm}\delta_{kl} \tag{2.29}$$

where ϵ_{ijk} is the Levi-Civita tensor and δ_{ij} is the Kronecker delta. This is useful for evaluating triple cross-products such as the BAC-CAB rule:

$$\vec{A} \times (\vec{B} \times \vec{C}) = \vec{B}(\vec{A} \cdot \vec{C}) - \vec{C}(\vec{A} \cdot \vec{B}). \tag{2.30}$$

Problem 5: Rotations and Matrix Multiplication

1. Consider the vector $\vec{A} = (2, -1, 5)$. Find $\vec{A}'$ in a frame rotated about the y axis by 45°.

2. Consider the vector $\vec{A} = (3, -2, 1)$. Find $\vec{A}'$ in a frame rotated about first the z axis and then the x axis, in each case by 45°.

3. Consider the 4-vector $A^\mu = (3, 1, -2, 1)$. Find $A^{\mu'}$ in a frame rotated around the z axis by 45° and then boosted along the y axis by velocity β.

4. Calculate the invariant length $|A^\mu|^2$ before and after the transformations in Parts 1–3.

Problem 6: Addition of Velocities in the Same Direction

1. A rocket ship moving at $\beta = 0.995$ away from Earth launches a small ship in the same direction with $\beta = 0.995$ relative to the mother ship. Find the velocity measured from Earth.

2. A rocket ship approaching Earth at $\beta = 0.9999$ turns on a headlight pointed right at Earth. Use the law of addition of velocities to calculate the velocity of the light from the headlight as seen on Earth.

3. A bottle is thrown backward from a moving car. If the car is moving at 90 feet/second and the bottle is thrown at 60 feet/second, how fast is the bottle moving with respect to the ground?

Problem 7: Addition of Velocities in Different Directions

Consider a frame F moving in the x direction at velocity β_1 relative to a frame F', which is itself moving at velocity β_2 *in the y direction* relative to a frame F''. Find the Lorentz transformation from F to F''.

Challenge Problem 8: Motivating the Lorentz Transformation Matrix

Transforming a history between moving frames of reference is often difficult for our non-relativistic visual-based intuition. However, determining the history in the moving frame is straightforward in the language of transformations using matrices; once set up correctly the solution follows algebraically. One can then ask what the time and place of the events are, interpreting what is seen taking into account light transit time. Here is the recipe:[19] (1) write the coordinates of events in frame F; (2) transform each event to frame F'; and (3) then visualize the history. Here we motivate the coefficients γ and $\beta\gamma$ in the algebra for a transformation from frame F to F' along the x axis, where F' is moving at velocity β in F.

1. Derive the coefficients of t and x in Eq. 2.24 (the top row of the matrix in equation 2.28) by transforming the spatial origin $(t, x = 0)$ in frame F to frame F' (use the result of Einstein Gedanken 1, i.e., the last line of Equation 1.1).

2. By equating the invariant length (squared) $|x^\mu|^2$ (Eq. 2.23) in Frames F' and F, show that the transformation of x to x' (the second row of the matrix in 2.28) satisfies the invariance of the length (squared) in Eq 2.23.

3. The inverse, M^{-1}, of the transformation matrix M of Equ. 2.28, is given by replacing the velocity β with $-\beta$. Show that the product $MM^{-1} = I$, where I is the identity matrix.

Challenge Problem 9: The Twin Paradox (An example of visual versus symbolic problem solving.)

Two identical twins flip a coin as to who gets to go on a trip to a distant exotic resort. The space cruise ship accelerates rapidly to $v = 2/3\ c$, goes for 1 year, and rapidly decelerates to land. After a brief visit, the process is reversed. Ignoring the time spent accelerating and visiting, the vacationing twin spent 2 years on the ship.

1. How many frames of reference are involved? Briefly describe each.

2. Are the travels of the twins really identical as implied?

3. Suppose a light flashes every New Year's Eve (midnight Dec 31, local time) on Earth and another flashes also on New Year's Eve (local time) on the space-ship. Draw a 2-dimensional *space-time diagram* (t vs. x as the vertical and horizontal axes, respectively) of the motion of the ship. Draw lines representing the transit of the light flashes from the Earth to the ship, and from the ship to Earth.

---

[19] I refer to this as "turning the crank" as it is very sure-footed.

4. Make a list of the times when the twin on the ship sees flashes from Earth.

5. Make a list of the times when the twin on Earth sees flashes from the ship.

6. If you enjoy programming, code this up in Python. However, it's instructive to draw it by hand first.

Finally, how much time has elapsed for the twin who remained on the Earth?

CHAPTER 3

Relativistic Kinematics; Energy and Momentum

3.1   Introduction

Our description of motion is based on the Principle of Relativity, which we have translated as the principle that in all inertial frames the laws of physics are expressed by the same equations. A corollary of the principle is that the quantities used in the description must transform from one frame to another in a way that preserves the invariance of the equations of motion. In Chapter 1 we have covered how time and position transform, which is as components of a 4-vector. Here we introduce energy and momentum, the conserved quantities that correspond to invariance under translations in time and space, respectively.

3.2   Conservation of Energy and Momentum

In addition to the invariance principle that these Laws of Physics are the same in all inertial frames, we postulate that these Laws are the same in all places and times in any given inertial frame.[1] This is called "invariance under translation" in space and in time, respectively.[2] Note that it applies to changing one or more of the four coordinates in a given frame, $t' = t + \Delta t$, or $x' = x + \Delta x$, rather than a transformation between two frames moving with respect to each other.

The postulated principle is that a translation in any of the 4 independent coordinates $t, \vec{x}$ leaves the laws of physics invariant. Each of the four invariances will result in a conserved quantity, with the four quantities energy, E, and 3-momentum, $\vec{p}$, being the conserved quantities associated with translations in time and space, respectively.[3]

---

[1] I confess I doubt that this has been true for all time, whatever that means. However, it has held up experimentally so far.

[2] Translation as in moving sideways, not as transcribing into another language.

[3] The relationship between invariance and conserved quantities is known as Noether's Theorem, after the exceptional mathematician Emmy Noether [13]. The proof is beyond the scope of this book and the author.

It is consequently natural that E and $\vec{p}$ are the components of a 4-vector. We write the 4-vector for a particle[4] as:

$$p^\mu = (E, cp_x, cp_y, cp_z).\tag{3.1}$$

3.2.1   Mass: The Total Energy E of a System in Its Rest Frame

Consider a particle in its rest frame. The momentum $\vec{p}$ is zero by definition, and so the 4-vector is given by:

$$p^\mu = (mc^2, 0, 0, 0)\tag{3.2}$$

where m is the mass.[5] The mass of an object is defined as the total energy of the system in its rest frame, including internal energies such as heat, and chemical, atomic, nuclear, and quark binding energies.[6]

In Classical Mechanics in principle we make a distinction between inertial mass, the mass that appears in Newton's Second Law $m\vec{a} = \vec{F}$, and gravitational mass, the mass that appears in Newton's Law of Gravitation, $\vec{F} = GmM/r^2$. However the two are measured to be equal, and are always blithely cancelled in solving mechanics problems.[7]

3.2.2   Units for Energy E, Momentum p, and Mass m

The only sensible units for relativistic calculations involving motion are Natural Units in which $c \equiv 1$. The ever-present appearance of c along with p as pc and with m as mc^2, and in β and γ as v/c, adds mind-numbing clutter, slows down calculating, and has no content beyond a choice of units. The rule to recover SI units is simple: replace t with ct, p with pc, m with mc^2, and replace v with v/c if v is dimensionless as in β and γ. You can then use meters and seconds instead of feet and nanoseconds, with $c = 3.0 \times 10^8$ m/s rather than 1.0 foot/nsec.

Energy has the dimension of mass times velocity squared. The units of energy in SI units are Joules, where 1 J is $1\ kg - m^2/s^2$. For atomic, nuclear, and elementary particle physics a Joule is much too large[8] compared to the energy levels involved; practitioners in these fields typically use electron volts (eV). One electron volt is the change in energy for an electron (or any other singly-charged particle, e.g., a proton) crossing a potential difference of 1 Volt. One thousand eV is one keV (note the small k); a million (10^6) eV is one MeV (Mega, large M); a billion (10^9) eV is

---

[4] We use the word particle for a small object for which we will keep track of only the the time and position, and the energy and momentum. We will discuss systems of particles in Chapter 4 and the motion of extended bodies in Chapter 8.

[5] In a particularly offensive usage, this is often called the "rest mass." There is no other kind in this context—see Okun [14] for a highly indignant reaction.

[6] Unnervingly, quark binding energies make up more than 98% of the mass of our complacently-described "ordinary" matter.

[7] In Galileo's famous (Gedanken?) dropping of objects off of the Leaning Tower of Pisa, cancelling the two masses in $m_i a = m_g g$, a cannon ball would fall with acceleration $a = g \approx 10 m/s^2$.

[8] One Joule is approximately the energy needed to lift a quart of Half-and-Half the height of a coffee mug.

one GeV (G for Giga); and 10^{12} eV is one TeV (T for Tera). Atomic energy levels are typically in the range of eV (outer electrons) to 10s of keV (inner electrons); nuclear levels are in the MeV range. The proton mass is 0.938 GeV, which we take here to be 1 GeV as good enough for folk-singing.

In SI units momentum has dimensionality mass×velocity, and consequently the dimensionalities of energy and momentum, and momentum and mass, differ by one power of velocity. Natural Units (NU) are more natural (*sic*). Since $c = 1$ in NU, energy, momentum, and mass are all measured in electron volts, eV.[9] We will work in NU for solving relativistic problems.

3.2.3   Relationships of Energy E and Momentum p to Mass m for a Particle

From Eq. 3.1, the momentum 4-vector in the particle rest frame (F) is

$$p^\mu = (m, 0, 0, 0). \tag{3.3}$$

Transforming p^μ from the rest frame to a moving frame F':

$$\begin{pmatrix} E' \\ p'_x \\ p'_y \\ p'_z \end{pmatrix} = \begin{pmatrix} \gamma & \beta\gamma & 0 & 0 \\ \beta\gamma & \gamma & 0 & 0 \\ 0 & 0 & 1 & 0 \\ 0 & 0 & 0 & 1 \end{pmatrix} \begin{pmatrix} m \\ 0 \\ 0 \\ 0 \end{pmatrix}, \tag{3.4}$$

i.e., reading off each row top to bottom:

$$\begin{aligned} E' &= \gamma m \\ p'_x &= \beta\gamma m. \end{aligned} \tag{3.5}$$

For convenience, we can drop the primes without any loss of generality:

$$E = \gamma m \quad \vec{p} = \vec{\beta}\gamma m. \tag{3.6}$$

3.3   The Invariant Length of p^μ

In the convention for 4-vectors of Chapter 1, the invariant length (squared) of the energy-momentum 4-vector p^μ is defined as:

$$|p^\mu|^2 \equiv E^2 - |\vec{p}|^2 \tag{3.7}$$

---

[9] To convert back to SI units, put the factors of c back in: momentum becomes pc and mass becomes mc^2. And, do not forget that in SI $c = 3 \times 10^8 \ m/s$.

where E is the energy. Writing the magnitude of the momentum generically without the prime or vector notation as p and the energy as E,

$$|p^\mu|^2 \equiv E^2 - p^2. \tag{3.8}$$

3.4   The "Master" Relationship $E^2 = p^2 + m^2$

Because Eq. 3.8 is the invariant length of a 4-vector, the value is the same in all frames. Evaluating the invariant length both in the rest frame and the lab frame gives us the relationship between E, p, and m.

In the rest frame $p^\mu = (m, 0, 0, 0)$, and so $|p^\mu|^2 = m^2$.

In the lab frame, $|p^\mu|^2 = E^2 - p^2$.

Equating the expressions in the two frames yields the "master equation" in an elegant way:

$$m^2 = E^2 - p^2$$
$$E^2 = p^2 + m^2. \tag{3.9}$$

Note that if the particle is at rest in the lab frame, $\vec{p} = 0$ and the equation for energy is $E = m$. In SI units the equation with $\vec{p} = 0$ reads $E = mc^2$, and for $p \neq 0$ reads $E^2 = (mc^2)^2 + (pc)^2$. I recommend sticking with Natural Units, and putting the factors of $c = 3.0 \times 10^8$ m/s back at the end if you need an answer in meters, kilograms, and seconds.

3.4.1   Photons: Energy and Momentum

Photons are massless, i.e., $m = 0$. From the above master equation, $E^2 = p^2 + m^2$, and so for photons $E = p$.[10] Light carries momentum. Note that the momentum of the photon is a 3-vector, carrying momentum in the direction of motion. The magnitude of the momentum is equal to the photon energy.[11]

3.4.2   Summary and Comment on Memorization

This course is by design light on memorization. However, having just derived the key formulas relating mass, energy, momentum, and velocity, we summarize what you should know cold.

We assume the particle or object is moving with velocity β in our frame, and replace p_x' with p for convenience. We use NU in the following summary

---

[10] Remember we are using NU here, and so both E and p are measured in eV and are numerically equal. In SI units, $E = pc$, measured in Joules.

[11] This is another benefit of Natural Units; one can use MeV or GeV for mass, momentum, and energy, rather than clumsily using MeV/c or GeV/c for momentum and MeV/c² or GeV/c² for mass everywhere they appear.

below; to convert to SI, replace m with mc^2, p with pc, where m is in kg, and $c = 3 \times 10^8$ m/sec.

1. Finding E and p given β and or γ:

$$E = \gamma m$$
$$p = \beta \gamma m. \tag{3.10}$$

2. Finding β and or γ given E and p:

$$\beta = p/E$$
$$\gamma = E/m. \tag{3.11}$$

For example, a 6 TeV 10^{12} eV proton in the CERN Large Hadron Collider (LHC) has $\gamma = E/m = 6000/1 = 6000$, and $\beta = \sqrt{((6000)^2 - 1)/(6000)^2} \approx 0.999999986$.

3. The 'master equation' relating E, p, and m:

$$E^2 = p^2 + m^2 \tag{3.12}$$

3.5   Problem Set 3: Energy and Momentum; Relativistic Kinematics

Time Management and Study Groups: Please don't be put off by the problems. They are not as hard as they may seem at first glance—you have very powerful tools. However, you really truly must work with your study group. Remember the work you hand in **has to be your own**.

The goal is for *all* students to do well. It is not competitive, and there should be no weeding out.[12]

Formulae:

$$\beta = v/c; \quad \gamma^2 = 1/(1 - \beta^2); \quad \beta^2 = (\gamma^2 - 1)/\gamma^2.$$

The invariant length of the 4-vector x_0, x_1, x_2, x_3: $|x^\mu| = \sqrt{x_0^2 - x_1^2 - x_2^2 - x_3^2}$.

---

[12] The mantra is "I am here to teach you, not to grade you. However, grading is an essential part of teaching." In other words, relax.

Lorentz transformation: a boost of an event in the x direction in frame F into frame F':

$$t' = \gamma t + \beta \gamma x \qquad (3.13)$$

$$x' = \beta \gamma t + \gamma x \qquad (3.14)$$

$$\begin{pmatrix} t' \\ x' \\ y' \\ z' \end{pmatrix} = \begin{pmatrix} \gamma & \beta\gamma & 0 & 0 \\ \beta\gamma & \gamma & 0 & 0 \\ 0 & 0 & 1 & 0 \\ 0 & 0 & 0 & 1 \end{pmatrix} \begin{pmatrix} t \\ x \\ y \\ z \end{pmatrix}. \qquad (3.17)$$

$$y' = y \qquad (3.15)$$

$$z' = z \qquad (3.16)$$

Invariant length of x^μ: $|x^\mu|^2 = t^2 - x^2 - y^2 - z^2$.

Energy-momentum 4-vector: $p^\mu = (E, p_x, p_y, p_z) = (E, \vec{p})$.

Invariant length of p^μ: $|p^\mu|^2 = E^2 - \vec{p}^2$.

Problem 1: Relationship of E, $\vec{p}$, and m for a particle

Consider a proton (yes, just one) accelerated at Fermilab up to an energy of 120 GeV. The proton mass is 0.938 GeV; you can approximate it as 1.0 GeV.

1. Draw a good, clear, labeled diagram of the problem and the two frames (proton rest frame and lab frame).

2. Find the Lorentz factor γ of the proton.

3. Find the velocity β of the proton.

4. Starting with the 4-vector in the proton frame, find the 4-vector in the Fermilab frame.

5. Find the invariant length of p^μ in both frames.

Problem 2: An Alternative Derivation of the Master Relationship

Starting with $E = \gamma m$ and $p = \beta \gamma m$, derive the Master Relationship:

$$E^2 = p^2 + m^2. \qquad (3.18)$$

Problem 3: Deriving the Astronomical Red Shift

Consider a very distant galaxy in our expanding Universe. A Lyman-α ($\lambda = 1216$ Angstroms,[13] corresponding to the $n = 2$ to $n = 1$ transition in Hydrogen) photon from the galaxy arrives at Earth. Other measurements indicate that the galaxy is receding from Earth at a velocity β.

---

[13] An Angstrom is $10^{-10} m$.

1. Draw a good clear labeled diagram of the two frames. Show the photon direction.

2. Derive the formula for the shift in energy of the photon starting with the 4-vector in the galaxy frame and transforming it into the Earth's frame.

3. The line in the spectrum from the galaxy is measured to be at 6000 Angstroms; find β for the galaxy.

Problem 4: The Kinematics of Higgs Boson Decay

A Higgs boson[14] is created at rest in a proton-proton collision at the LHC. The Higgs decays into a b-quark and an anti-b-quark ($b\bar{b}$). The mass of the Higgs is M_H; the mass of a b-quark or a $\bar{b}$ anti-quark is m_b. Solve for the magnitude of the momentum of the b-quark in terms of the masses. (Be kind to your grader—show every step, and line up the equal signs neatly, one under the next.)

Problem 5: The Discovery of the Top Quark

A top quark decays to a W boson and a bottom quark. The W boson then itself decays into an electron and neutrino. In the Collider Detector at Fermilab (CDF experiment) a group of UC undergrads could measure the 3-momenta of the bottom quark (b-quark), the electron, and the neutrino. We know the masses of each. Writing the 4 vectors for the b-quark, the electron, and the neutrino as (E_b, px_b, py_b, pz_b), (E_e, px_e, py_e, pz_e) and $(E_\nu, px_\nu, py_\nu, pz_\nu)$, respectively, write down the formula for the mass of the top quark (this is what we did to discover the top).

Problem 6: Electron-Positron Annihilation for Medical Imaging

Positron-emission tomography (PET) is a medical diagnostic used to detect hairline fractures, cancer, and other hot-spots of biological activity. The positron is the anti-electron, identical except positively charged. A positron and an electron annihilate to two gamma rays (energetic photons). A radioactive tracer such as ^{18}F is ingested and migrates to sites of activity such as tumors or breaks. The gamma rays can be detected and linked to the site of the positron annihilation. Given the mass of the electron to be 0.511 MeV, solve for the energy and momenta (be careful) of the gammas.

Problem 7: The Taylor Expansion as an Everyday Tool

Use a Taylor expansion in the appropriate small quantity to approximate the following when the small quantity is small. When the number of terms is not explicitly asked for, do enough so that the series is clear.

---

[14] Instructive in that to solve this you do not need to know what a Higgs boson or a b-quark is, but only that the Higgs spontaneously decays into a b-quark and an anti-b-quark, and that the masses of the latter are equal. However, *Grace in All Simplicity* by R. Cahn and C. Quigg in the recommended reading contains an excellent personal exposition on the Higgs and the heavy b- and c- quarks, and, for Problem 5, the top quark.

1. $(1+x)^n$ to fourth order in x ($\mathcal{O}(x^4)$), for $x << 1$.

2. $(1.02)^6$.

3. $\sqrt{0.96}$.

4. $(0.96)^{-1/3}$.

5. The sum of two co-linear small velocities: $\beta_T = \frac{\beta_1 + \beta_2}{1 + \beta_1 \beta_2}$.

Problem 8: Newtonian Approximations for Energy and Momentum

The familiar Newtonian energy and momentum formulae can be derived as follows:

1. Expand the Lorentz factor γ in v/c to order $(v/c)^4$ for the non-relativistic case $v << c$;

2. For a particle of mass m, similarly expand the relativistic energy E to order $(v/c)^4$;

3. For a particle of mass m, expand the relativistic momentum p to order v/c;

4. For a 6-ounce baseball moving at 102 miles/hour, evaluate the first 3 terms of the Taylor expansion for the energy in SI units.

Problem 9: Systems of Particles

Consider a system of 3 relativistic particles each moving with respect to the lab frame. Calculate:

1. The energy of the system;

2. The momentum of the system;

3. The mass of the system;

4. The Lorentz factor of the system in the lab frame;

5. The velocity $\vec{\beta}$ of the system in the lab frame.

6. What is the velocity $\vec{v_{CM}}$ of the system if the particles are non-relativistic, e.g., billiard balls?

Challenge Problem 10: The Fermilab Tevatron and the CERN Large Hadron Collider (LHC)

1. For the Fermilab conditions given in Problem 1, calculate the energy in the center-of-mass frame of a collision of a proton in the 120 GeV beam and a proton at rest (i.e., in the lab frame).

2. Do the same for two 120 GeV beams colliding head-on.

3. The Large Hadron Collider (LHC) at CERN collides two 6.5 TeV (Tera-electron volt) beams head-on. Find the energy in the center-of-mass frame.

Challenge Problem 11: The Highest Energy Cosmic Rays and the GZK Cutoff (Not hard; just lots of history.)

The production mechanisms, distribution and location of sources, and composition of the highest energy cosmic rays arriving at Earth is a frontier.

In 1991 Jim Cronin and Alan Watson started the development of a ground array of cosmic ray detectors with the goal of a coverage of 1000 km^2, in what became the Pierre Auger Observatory, located in the Pampa Amarilla in the Mendoza Province of Argentina [35]. In 1966, Greisen [36], and Zatsepin and Kuzmin[37], had proposed the interaction of very high energy cosmic rays with the very low energy photons in the Cosmic Microwave Background (CMB). In 1952 Enrico Fermi and co-workers had discovered a short-lived excited state of the proton, the Δ (Delta), at a mass of 1238 MeV. At that energy in the center-of-momentum frame a cosmic ray proton colliding with a CMB photon would produce a Δ, which then decays to a pion and a lower energy proton, effectively removing the incident proton from the cosmic ray spectrum. The GZK mechanism, as it is known, thus predicts an upper limit, the GZK cutoff, to the ultra-high energy cosmic ray spectrum, but under the assumption that the incoming particles are protons.

Call the energy of the incident cosmic ray proton E_p, and the energy of the CMB photon E_γ. The CMB energy is characterized by the temperature of 2.7 degrees Kelvin, which corresponds to 2.33×10^{-4} eV. We want to solve for the value of E_p that produces a center-of-mass energy of 1238 MeV, the mass of the Δ.

1. Write the 4-vector for the photon in the Earth's frame.

2. Write the 4-vector for the proton in the Earth's frame (you can neglect the mass; 1 GeV is negligible).

3. Add the two 4-vectors component by component (write them one above the other and add the columns).

4. Calculate the invariant length of the summed 4-momentum for the system.

5. Equate the sum with 1238 MeV and solve for E_p, the energy of the proton needed to produce the Δ off of the CMB.

6. (Optional) If you are curious about the composition of the very highest charged cosmic rays, look up the present status of the GZK cutoff.

CHAPTER 4

The Non-Relativistic Limit $c \to \infty$ and Newtonian Mechanics

4.1   Introduction

Newtonian kinematics corresponds to the case of taking the limit $c \to \infty$, a very different universe from the one we live in. In this limit the dimensionless velocity $\beta \equiv v/c \to 0$ in the relativistic formulas for energy, momentum, and transformations from one inertial frame of reference to another.

To derive Newtonian kinematics,[1] we expand the kinetic energy E and the momentum p in a Taylor expansion[2] in the small quantity v/c. We will convert to SI units, and so put the factors of $c \equiv 3.0 \times 10^8$ m/s back in where they have been silent partners. Lastly, we will ignore entirely[3] the *largest* term in the energy, mc^2, and higher-order terms in v/c than $(v/c)^2$ in the energy and v/c in the momentum. Note that the Lorentz factor γ becomes identically 1. The transition to Newtonian kinematics takes only three or four lines.[4]

For non-relativistic problems we will work in SI (MKS) units.

4.2   Expanding γ in Powers of the Small Quantity v/c

For velocities $\beta << 1$, i.e., $v << c$, a Taylor expansion of the Lorentz factor γ gives:

$$\gamma \equiv \frac{1}{\sqrt{1 - v^2/c^2}} \equiv (1 - (v/c)^2)^{-1/2} \approx 1 + \frac{1}{2}(v/c)^2 + \frac{3}{8}(v/c)^4 + \ldots \quad (4.1)$$

---

[1] I assume you have been introduced to Newton's 3 Laws in a course in AP Physics or equivalent. The three 17th-century ideas are utterly brilliant.

[2] See Appendix A, Section A.1.3.

[3] This is known affectionately among some as "The Atomic Bomb Approximation."

[4] And you cannot go in the inverse direction at all.

4.3   Kinetic Energy for $v << c$: $T = \frac{1}{2}mv^2$

To find the kinetic energy of an object of mass m moving at velocity $v << c$ we expand the energy $E = \gamma mc^2$ in a Taylor series:

$$E = \gamma mc^2$$

$$\approx mc^2(1 + \frac{1}{2}(v/c)^2 + \frac{3}{8}(v/c)^4 + \ldots) \tag{4.2}$$

$$= mc^2 + \frac{1}{2}mv^2 + \frac{3}{8}mv^2(v/c)^2 + \ldots$$

The first term, mc^2, is the energy of the object at rest,[5] and is much larger than the kinetic energy. However in classical mechanics by definition we assume that matter is neither created nor destroyed, and so the rest energies are unchanging and can be ignored in calculating changes in energy. The third and subsequent infinite number of terms we also ignore, as they are at least of order $(v/c)^2$ smaller than the Newtonian definition of $E_{kinetic} \equiv \frac{1}{2}mv^2$, where m is the mass in kg and v is the velocity in m/s.

Using only the 2nd term in the expansion, we have the expression for the non-relativistic kinetic energy for a particle of mass m:

$$T \equiv \frac{1}{2}mv^2. \tag{4.3}$$

4.4   Momentum: $\vec{p} = m\vec{v}$

We start from the relativistic formula for momentum, $p = \beta \gamma m$, where m is the mass. Working in SI units, we write m as mc^2, β as v/c, and expand γ in powers of the small quantity v/c:

$$p = \beta \gamma m$$

$$pc = (v/c)(mc^2)\gamma$$

$$p = (v)(m)\gamma \tag{4.4}$$

$$p = mv(1 + \frac{1}{2}(v/c)^2 + \frac{3}{8}(v/c)^4 + \ldots).$$

Using only the first term in the expansion, and treating each component independently, we have the expression for the non-relativistic momentum

$$\vec{p} \equiv m\vec{v}. \tag{4.5}$$

---

[5] Also known as the mass.

Note that the correction to the non-relativistic approximation is order $(v/c)^2$.

To see how small this is, consider a car traveling at 60 mph (miles/hour). Using our method of converting units by multiplying by the expression for 1 (unity) that replaces the unit we want to get rid of with the one we want:

$$
\begin{aligned}
v &= 60 \text{ miles/hour} \\
v &= 60 \text{ miles/hour} \times (5280 \text{ feet/mile}) \times (1 \text{ hour/3600 s}) \\
v &= 88 \text{ ft/sec} \times 10^{-9} \text{ sec/nsec} \\
v &= 88 \times 10^{-9} \text{ feet/nsec} \\
v &= 88 \times 10^{-9} \text{ c}
\end{aligned}
\tag{4.6}
$$

$$(v/c)^2 \approx 8 \times 10^{-15}$$

4.5   Non-Relativistic Addition of Velocities: The Galilean Transformation

In Section 2.6.1 we found the relativistic equation for the total velocity resulting from the addition of co-linear velocities β_1 and β_2:

$$\beta_T = \frac{\beta_1 + \beta_2}{1 + \beta_1\beta_2}. \tag{4.7}$$

Working in SI units and putting in the factors of $1/c$, where $c = 3 \times 10^8$ m/s:

$$v_T/c = \frac{v_1/c + v_2/c}{1 + (v_1 v_2)/c^2}. \tag{4.8}$$

Invoking the Taylor expansion to lowest order,

$$(1+x)^{-1} \approx (1-x) \tag{4.9}$$

and cancelling the ever-present c's, we recover the non-relativistic approximation for the addition of velocities:

$$
\begin{aligned}
v_T/c &= \frac{v_1/c + v_2/c}{(1 + (v_1 v_2)/c^2)} \\
v_T &= \frac{v_1 + v_2}{(1 + (v_1 v_2)/c^2)} \\
v_T &= (v_1 + v_2)\,(1 + (v_1 v_2)/c^2)^{-1} \\
v_T &\approx (v_1 + v_2)\,(1 - (v_1 v_2)/c^2 + \ldots) \\
v_T &\approx v_1 + v_2.
\end{aligned}
\tag{4.10}
$$

Taking the limit $c \to \infty$, and generalizing to all 3 dimensions:

$$\vec{v_T} = \vec{v_1} + \vec{v_2}. \tag{4.11}$$

This (approximate!) law of addition of velocities is called the Galilean Transformation after Galileo Galilei. It is how we transform between frames in our daily life. Note that the relativistic correction is of order $(v_1 v_2)/c^2$.

4.6   Systems of Particles and the Center-of-Momentum Frame

The non-relativistic formula for the velocity of the center-of-momentum (c.m.) system directly follows from the relativistic formula for the velocity of a particle, $\beta \equiv p/E$.

Consider a system of N particles moving in frame F. Summing the components of the 4 momenta individually (column-by-column) to get the total momentum p_T^μ:

$$
\begin{aligned}
p_1^\mu &= (E_1, & p_{1x}, & \quad p_{1y}, & \quad p_{1z}) \\
p_2^\mu &= (E_2, & p_{2x}, & \quad p_{2y}, & \quad p_{2z}) \\
&\;\vdots \\
p_N^\mu &= (E_N, & p_{Nx}, & \quad p_{Ny}, & \quad p_{Nz})
\end{aligned}
\tag{4.12}
$$

Summing by columns

$$
\begin{aligned}
p_T^\mu &= ((E_1 + E_2 + \ldots), \; (p_{1x} + p_{2x} + \ldots), \; (p_{1y} + p_{2y} + \ldots), \\
& \quad (p_{1z} + p_{2z} + \ldots))
\end{aligned}
$$

The velocity of the center-of-mass frame in the lab frame is given by

$$\vec{\beta} = \frac{\vec{p_T}}{E_T}. \tag{4.13}$$

In the non-relativistic limit:

$$\vec{V}_{CM} = \frac{\sum_{i=n}^{N} m \vec{v_i}}{\sum_{i=n}^{N} m_i}, \tag{4.14}$$

i.e., the velocity of the center-of-mass of the particles is the mass-weighted average of the velocities.[6]

---

[6] In the non-relativistic case "center-of-mass" and "center-of-momentum" tend to be used interchangeably, often ambiguously labeled as CM. In the relativistic case we often refer to center-of-mass when we mean center-of-momentum (easier to say, perhaps?).

4.7   Non-Relativistic (NR) Collisions

We have already been dealing with "before-after" problems such as the decays of the Higgs boson and the top quark, or positron-electron annihilation in medical imaging. We used the principles of conservation of energy and momentum to predict the behavior of a system after a sudden change from one well-defined state to another.

The non-relativistic case of motion of a particle is similar in principle; however, there is a significant difference in the meaning of the symbol for energy, E. Up to this point we have used E to be the total energy of the particle, where $E^2 = p^2 + m^2$ (in NU), and the mass of the particle includes the energies in internal degrees of freedom, such as binding energies. In Newtonian mechanics we instead will use the non-relativistic (NR) expression for energy $E = T + V$, where $T = \frac{1}{2}mv^2$ is the kinetic energy and V(r) is the potential energy. The relativistic E is *always* conserved; the non-relativistic E by convention is not, as the definition does not include other forms of energy such as heat.

Collisions form the most common non-relativistic "before-and-after" problems. We will ignore the interaction forces during the collision, and make the assumption that the system has well-defined energy and momentum both before and after the collision. Collisions are classified as elastic if E is conserved,[7] and inelastic if not. The most common cases of inelastic collisions (e.g., car crashes) involve the loss of kinetic energy into heat and/or the deformation of materials.

4.7.1   Collisions in Laboratory and Center-of-Mass Frames

A theme of the course from the beginning has been the power of using transformations between frames as a simplifying tool. The two frames most useful for analyzing non-relativistic collisions are the laboratory frame, in which one particle, the target, is at rest, and the center-of-momentum frame, in which there is no net momentum.

4.7.2   Transforming from the Lab Frame to the CM Frame
        or Vice Versa

We have seen in Section 4.5 that the Lorentz transformation from one frame to another in the NR limit reduces to the Galilean transform, which is just a simple addition of the relative velocity of the two frames to the velocities of each of the objects in the frame. Thus to go from the CM frame (for example Casals' frame) to the lab frame (Primrose's frame) in the non-relativistic limit, we

---

[7] Since $E = T + V$ and $V = V(r)$ is the same before and after, conservation of E is equivalent to conservation of T.

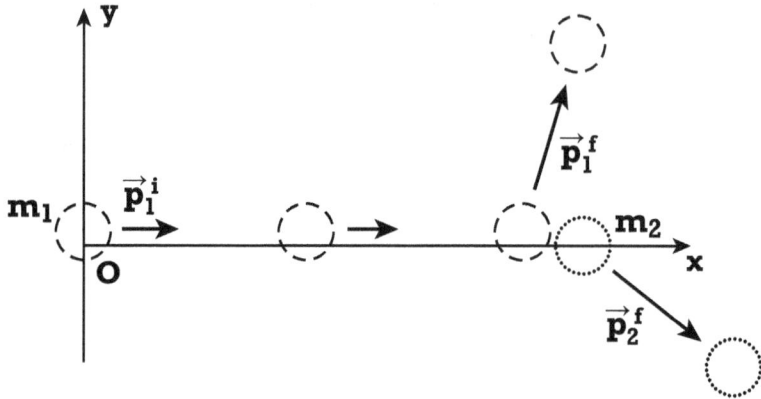

Figure 4.1. Elastic scattering. Particle 1, traveling along the positive x-axis, strikes particle 2 which is at rest. The respective masses are m_1 and m_2. We denote initial and final momenta as p_i and p_f. Since point masses are unphysical and a head-on collision would be purely one-dimensional, the particles are drawn as finite and slightly non-collinear. Ignore this when solving.

merely add the relative velocity $\vec{V}_{CM}$ (Equation 4.14) to all velocities in the CM frame.[8]

$$\vec{v}_{lab} = \vec{v}_{cm} + \vec{V}_{CM} \tag{4.15}$$

4.8   Elastic Collisions

In an elastic collision the kinetic energy T and the total energy E of the system are conserved, i.e. are the same before and after the collision.

4.8.1   Setting up a 2-Body Elastic Scattering Problem

We will work in the lab frame, as shown in Fig 4.1. The scattering will lie in a plane, which we will take as the xy plane. Particle 1, traveling along the positive x-axis, strikes particle 2 which is at rest. The masses are m_1 and m_2. We denote initial and final momenta as p^i and p^f.

The scattering is described by four unknowns: the magnitude and angle of p_1^f and p_2^f, respectively, as shown in Fig. 4.1.

To determine these 4 parameters we have only 3 constraints: 1) conservation of kinetic energy T (remember the collision is assumed to be elastic); 2) conservation

---

[8] Note that since we add $\vec{V}_{CM}$ to all vectors the transformation preserves the relative velocities of all objects in the transformation. The NR transformation is just a constant shift from one frame to another in what one defines as "at rest."

of momentum in the x direction; and 3) conservation of momentum in the y direction. We consequently cannot determine a unique solution—the scattering can take place for any outgoing angle of particle 1, for example. However, given one of the 4 unknowns we can solve for the other 3.

The constraint equations are:

$$\text{Conservation of T} \quad (p_1^i)^2 = (p_1^f)^2 + (p_2^f)^2$$

$$\text{Conservation of } p_x \quad p_{1x}^i = p_{1x}^f + p_{2x}^f \tag{4.16}$$

$$\text{Conservation of } p_y \quad p_{1y}^i = p_{1y}^f + p_{2y}^f.$$

Solving these three equations for the general solution is not hard but is tedious, and the solution is remarkable only in the almost-complete lack of opportunities to use it.[9]

4.8.2   Example 1: Equal Masses

There is a special case that is useful in practice and is simple algebraically when the masses of the colliding particles are equal, as in billiards or pool. In this case the two outgoing momenta form an angle of 90°.[10] The proof is sweet:

Conservation of kinetic energy $T = p^2/2m$:

$$(p_1^i)^2 = (p_1^f)^2 + (p_2^f)^2 \tag{4.17}$$

Conservation of momentum $\vec{p}$:

$$\vec{p}_1^i = \vec{p}_1^f + \vec{p}_2^f \tag{4.18}$$

Squaring equation 4.18 and subtracting equation 4.17 gives:

$$\vec{p}_1^f \cdot \vec{p}_2^f = 0 \tag{4.19}$$

i.e. the opening angle of the outgoing particles is 90°.

4.8.3   Example 2: Head-On Collision with a Particle at Rest

Consider a particle of mass M moving at velocity V. The particle collides elastically head-on with a particle of mass m initially at rest. By symmetry all motion is in one-dimension.

---

[9] You can look up the algebra on the web or in standard texts. In practice, in scattering experiments in particle or nuclear physics one usually has the necessary additional (4th) constraint of a measurement of either an outgoing angle or momentum.

[10] Except in the case of exactly head-on collisions where the angle is undefined.

We can solve this easily using conservation of kinetic energy T and momentum p, but there are times—such as on a standardized multiple-choice exam—where time is limited. It is much faster to use some vague memory of how the answer goes supplemented by dimensional analysis and limiting cases.

Let's call the outgoing velocities v_1 and v_2, where particle 1 is the incoming particle with mass M. Let's also make the educated guess (ansatz is the 50-cent word) that v_1 and v_2 are linearly proportional to V and depend only on M and m, i.e.,

$$v_1 = C_1(m, M)V \tag{4.20}$$

$$v_2 = C_2(m, M)V. \tag{4.21}$$

Consider three limiting cases:[11]

1. $M >> m$: $v_1 = V$, $v_2 = 2V$;

2. $M = m$: $v_1 = 0$, $v_2 = V$;

3. $M << m$: $v_1 = -V$, $v_2 = 0$ (lab and CM frames are identical).

Let's make educated guesses of dimensionless choices linear in the masses for C_1 and C_2 that satisfy these limits:

$$v_1 = \frac{M - m}{M + m}V \tag{4.22}$$

$$v_2 = \frac{2M}{M + m}V. \tag{4.23}$$

Letting $M \to \infty$, $M \to 0$, and $M = m$ gives the correct limits in the 3 cases (try it).[12]

4.8.4   Inelastic Collisions

In an inelastic collision kinetic energy T is not conserved, with energy being converted to internal degrees of freedom such as heat or mechanical distortion. However, the number of unknowns is reduced from the elastic case, and conservation of momentum provides enough constraints to solve for the motion after the collision.

We should (again) make clear that we believe[13] that energy and momentum are *always* conserved. Momentum completely contained in internal degrees of freedom cancels out pairwise due to Newton's Thrid Law (see Section 5.3), as momentum is vectorial. In contrast, energy in internal degrees-of-freedom is a scalar and is

---

[11] To visualize the first two cases, in your mind try transforming to the CM, do the scattering, and then transform back to the lab.

[12] This may seem cavalier, but dimensional analysis plus solving limiting cases is a key skill for taking multiple choice exams, to pick a particularly eGREgious example, and has a lot of merit in its own right.

[13] We-all, not the Royal We. The belief is based on data (measurements); no violations have yet been observed.

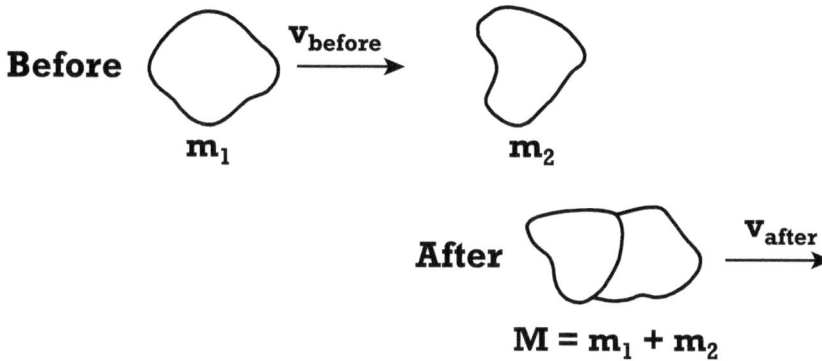

Figure 4.2. An inelastic 2-body collision. Particle 1 is traveling with velocity v_{before} and impinges on particle 2 which is at rest. After the collision the two move together as one object with velocity v_{after}.

additive, and so heat—the energy of atoms and molecules, for example—does not cancel pairwise, and so the energy converted to heat is accounted for as being "lost."

Figure 4.2 shows an inelastic 2-body collision. Particle 1 is traveling with velocity v_{before} and impinges on particle 2 which is at rest. After the collision the two move together as one object with velocity v_{after}.

Solving for v using conservation of momentum:

$$m_1 v_{before} = (m_1 + m_2) v_{after}$$

$$v_{after} = \frac{m_1}{m_1 + m_2} v_{before}. \tag{4.24}$$

Checking dimensions: the coefficient between the 2 velocities is dimensionless as it must be.

Checking limits: for $m_1 >> m_2$, $v_{after} = v_{before}$; for $m_1 << m_2$, $v_{after} = 0$; and for $m_1 = m_2$ $v_{after} = \frac{1}{2} v_{before}$, as it should be.

4.9   Problem Set 4: Newtonian Kinematics

Time Management and Study Groups: You truly **must** do this with your study group. The problem sets will go (at least) two or three times faster if you discuss the problems with friends/colleagues, and many puzzling concepts become your own by trading questions and explanations.[14] If your whole group is stuck, go find the Instructor and/or the Teaching Assistants. **However, the work you hand in has to be your own.**

---

[14] On a recent Teaching Evaluation form, a student responded to the question, "What was the most important thing you learned in this class?" by "I'm not in high school anymore."

Problems: Please use MKS (SI) units where appropriate. Take the acceleration of gravity on the Earth's surface to be g = 9.8 m/s^2, which we will often approximate as 10 m/s^2 when it doesn't matter.

Some of this material duplicates Problem Sets 2 and 3. It is that important and should be more than familiar. Your study group should polish it off quickly. It makes good quiz material.

Problem 1: Newtonian Approximations for Energy and Momentum

The familiar Newtonian energy and momentum formulas can be derived as follows:

1. Expand the Lorentz factor γ in v/c to order $(v/c)^4$ for the non-relativistic case $v \ll c$;

2. For a particle of mass m, similarly expand the relativistic energy E to order $(v/c)^4$;

3. For a particle of mass m, expand the relativistic momentum p to order v/c;

4. For a 6-ounce baseball moving at 102 miles/hour, evaluate the first 3 terms of the Taylor expansion for the energy in SI units.

Problem 2: The Non-relativistic Limit of the Addition of Velocities

Starting with the correct expression for the addition of two velocities in the same direction:

$$\beta_T = (\beta_1 + \beta_2)/(1 + \beta_1\beta_2) \tag{4.25}$$

1. Derive the "Galilean Transformation" $v_T = v_1 + v_2$ for $\beta_1, \beta_2 \ll 1$.

2. Justify why this simple addition of velocities could be considered a transformation between frames.

Problem 3: Collisions with Equal Masses

Please draw a clear careful diagram.

1. **The power of vectors over trigonometry.**
   Consider a moving hockey puck colliding with an identical puck at rest on a very smooth ice rink. Using conservation of energy and momentum $\vec{p}$, show that the angle between the two puck trajectories after the collision is 90 degrees.

2. **A special case—neither vectors nor trigonometry.** Solve for the case in which the first puck hits the stationary puck head-on.

3. **Energy is always conserved, kinetic energy almost never.**

Suppose instead of bouncing off each other elastically the two hockey pucks stick together in an inelastic collision. Find the velocity of the pucks after the collision.

Problem 4: An Example of Galilean Transformation

A tennis ball is hit horizontally at the back of a truck moving directly away along the direction of the ball. The truck is going 30 mph; the ball is traveling at 60 mph. Find the velocity of the ball after it bounces off the back of the truck. (You might want to transform to the truck's frame and back.)[15]

Problem 5: Dimensional Analysis and Limits in Reconstructing Formulae

Consider an object of mass M moving at velocity V colliding head-on with an object of mass m at rest. By merely (*sic*) considering dimensional arguments, symmetry, and the limits $M = m$, $M >> m$, and $M << m$, write down the formulae for velocities v_M and v_m after the collision.

Challenge Problem 6: An Inelastic Collision

You drive a nail into a board by hitting it hard with a hammer. How hot does the nail get?[16]

---

[15] Thanks to Frank Merritt.

[16] You may not be used to having to give names and values to parameters in problems, as in high school everything is presented wrapped with a ribbon on a velvet-lined silver tray ("Consider a sphere of radius R"). Define the necessary parameters, and find and reference reasonable values for them. Research is like that.

CHAPTER 5

Non-Relativistic Dynamics:
Newton's Second Law, Force, Work

5.1 Introduction

Dynamics is the science of motion under external forces. We start with Newton's Second Law, which we translate from the Latin[1] as:

$$d\vec{p}/dt = \vec{F}_{total} \tag{5.1}$$

where $\vec{p}$ is the (3-vector) momentum of the object and $\vec{F}_{total}$ is the total force impressed on the object.

5.1.1 Defining the Force $\vec{F}$ on an Object with Mass m

Consider an object of mass m at rest. Apply a force, for example a steady push, a compressed spring, gravity (imagine an apple on a tree), or a spring breeze to the object.[2]

The object will accelerate:

$$m\vec{a} = \vec{F}_{Tot}$$
$$\text{where} \quad \vec{F}_{Tot} \equiv \Sigma \vec{F}_i. \tag{5.2}$$

More generally, the Second Law is:

$$d\vec{p}/dt = \Sigma \vec{F}_i \tag{5.3}$$

where $\vec{p}$ is the momentum of the object, $\vec{p} = m\vec{v}$.

---

[1] Mutationem motus proportionalem esse vi motrici impressae, et fieri secundum lineam rectam qua vis illa imprimitur.

[2] For now we neglect rotations and assume that the object is symmetrical (e.g., a small sphere) and the force is applied at the center of symmetry. We will deal with the more general case of torques and angular momentum in Chapters 6 and 7.

5.1.2   Force Is Proportional to the Time Derivative of Momentum Rather Than of Velocity

Equation 5.3 is the statement of Newton's Second Law in terms of the change in momentum caused by the impressed force rather than the more common but incomplete expression $\vec{F} = m\vec{a}$.

Specifically,

$$\vec{p} = m\vec{v}$$
$$d\vec{p}/dt = (dm/dt)\vec{v} + m(d\vec{v}/dt). \tag{5.4}$$

Given the limitations of a quarter/semester introductory course we will consider only cases for which $dm/dt = 0$.[3] The 2nd Law for constant mass is:

$$d\vec{p}/dt = m\vec{a}. \tag{5.5}$$

5.2   Solving Problems with Newton's Laws—Forces

Figure 5.1 shows three traditional pedagogical problems involving moving parts: an Atwoods Machine, a block on an inclined plane, and a Double Atwoods Machine. Each is idealized: we assume no friction between moving surfaces, all strings are massless, and pulleys are very small and have negligible moments of inertia.

5.2.1   Free-Body Diagrams

Consider the Atwood's Machine of Fig. 5.1. Each constituent part, such as each of the two weights or the string between them, obeys Newton's Second Law for the total force impressed on that part alone. A Free-Body Diagram (FBD) consists of a drawing of each part showing all individual forces on it. As the name implies, the body is treated as "free," i.e., constraints between bodies are ignored.

5.2.2   A Recipe

This genre of solving for the motion of a system of objects with inter-object constraints lends itself to a systematic common approach (a recipe):

1. Draw a generous clear diagram with a labeled coordinate system (axes, origin).

2. Make a Free-Body Diagram (FBD) for each object with mass that can move.

3. Write the Equations of Motion for each of the objects.

---

[3] For a clear treatment of problems in which $dm/dt \neq 0$, such as the acceleration of rocket-like objects, see Marion's Classical Dynamics. Try to get the elegant 1965 edition (Academic Press) if you can.

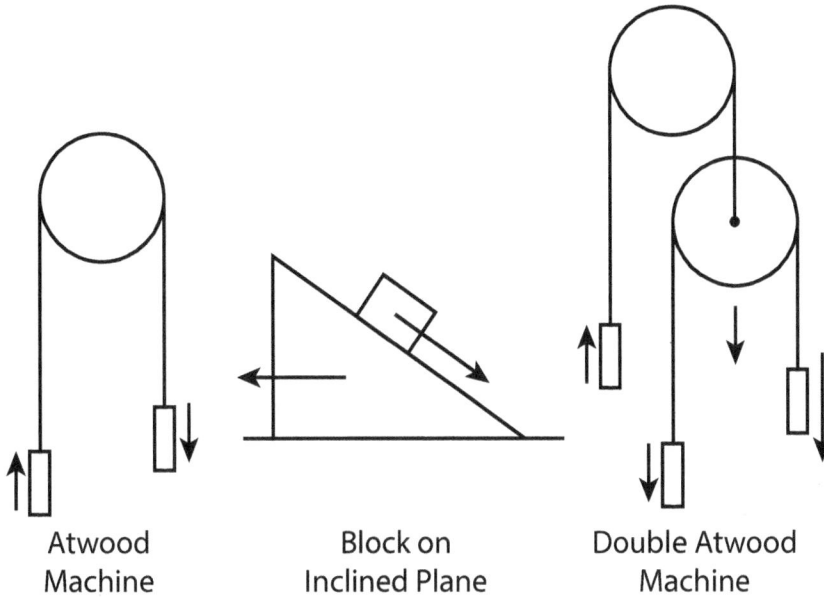

Figure 5.1. Three traditional pedagogical problems as examples for a systematic free-body-based solution: an Atwoods Machine, a block on an inclined plane sitting on a slippery horizontal surface, and a Double Atwoods Machine.

4. Count the number of Unknowns.

5. Write the Equations of Constraint.

6. Compare the number of Constraints to the number of Unknowns. If you do not have an equal number, stop here and think.

7. Solve the system of linear equations (you may want to use matrix methods).

8. Check limiting cases of the solutions and your intuition.

5.2.3   Example: The Atwood Machine

Following the recipe steps for the Atwood machine shown in Fig. 5.1:

1. **Draw a diagram.** The coordinates of the two masses in a Cartesian system are defined in Figure 5.2. We have chosen the signs such that both masses have positive motion defined to be in the positive y direction.[4] The rope is taken to be massless and of length L; the pulley is massless and is centered at a height $y = h$.

2. **Draw a Free-Body Diagram for each object.**

---

[4] The choice is arbitrary; one only has to be consistent throughout the calculation. The motion (of course) cannot depend on it.

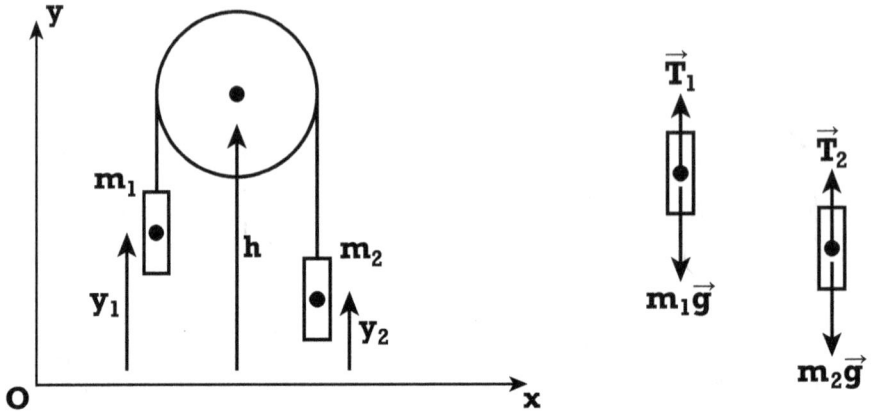

Figure 5.2. A diagram with origin, axes, and clearly defined labeled coordinates.

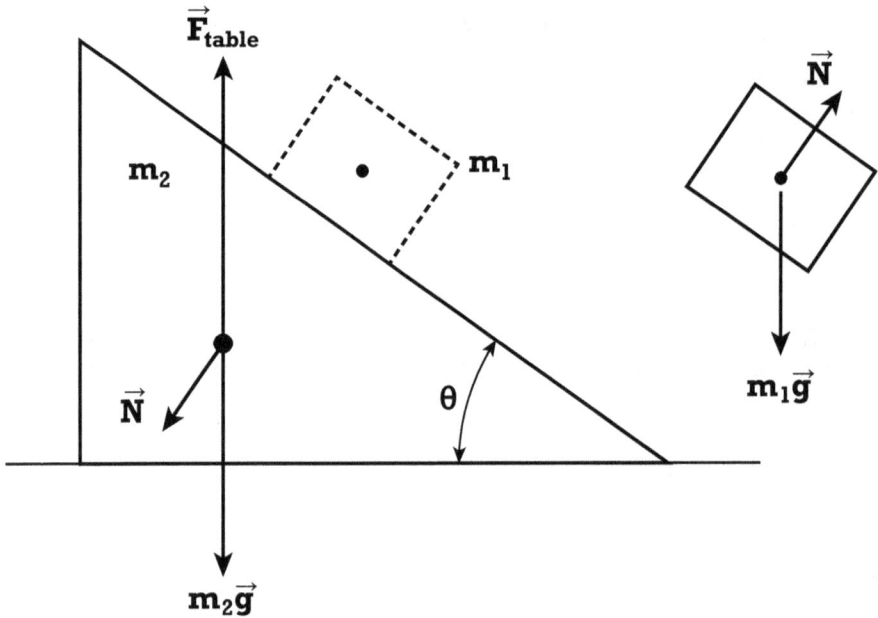

Figure 5.3. Free-Body Diagrams for the block sitting on a wedge that can slide.

Each of the masses has only two forces on it: gravity $m_1 g$ or $m_2 g$ in the negative y direction[5] and the tension T_1 or T_2 of the rope pulling in positive y.

3. **Equations of Motion.**
   We write the equations of motion for each of the two masses by reading off the forces from the respective FBDs, being careful to get the signs consistent with

---

[5] It's acceptable to use $g = 10$ m/s^2.

the positive direction of the coordinate system:

$$m_1 a_1 = T_1 - m_1 g$$
$$m_2 a_2 = T_2 - m_2 g. \tag{5.6}$$

4. **Equations of Constraint.**

The equations of constraint are the only part of the recipe that takes thought.

The rope is of length L. From the diagram, we can write L in terms of y_1, y_2, and the height of the (very small) pulley, h:

$$(h - y_1) + (h - y_2) = L$$
$$y_1 + y_2 = 2h - L. \tag{5.7}$$

We can get the accelerations of m_1 and m_2 by differentiating the coordinates with respect to time twice:

$$a_1 + a_2 = 0$$
$$a_1 = -a_2. \tag{5.8}$$

We can rename a_1 to be simply a, and so $a_2 = -a$. Note that when one mass goes up the other goes down, as it should.

There is one more equation of constraint, relating the tensions T_1 and T_2. Consider the general case where the rope has mass density λ kg/m. A short length $d\ell$ will have mass $dm = \lambda d\ell$. The Free-Body Diagram for this element of the rope is shown in Figure 5.4. Newton's Second Law gives

$$(dm)\, a = T_2 - T_1. \tag{5.9}$$

But $dm \to 0$ in the limit $\lambda \to 0$, and so $T_2 = T_1$. Note that we have proved the tension in a massless (i.e., light) rope is everywhere the same. We define $T \equiv T_1 = T_2$.

We now have 4 equations relating the 4 unknowns:

$$m_1\, a_1 = T_1 - m_1\, g$$
$$m_2\, a_2 = T_2 - m_2\, g$$
$$a_1 = -a_2 \equiv a \tag{5.10}$$
$$T_1 = T_2 \equiv T.$$

5. **Count Unknowns and Equations**.

Unknowns: a_1, a_2, T_1, T_2: i.e., 4 unknowns.
Equations: We have four equations, Eq. 5.10.

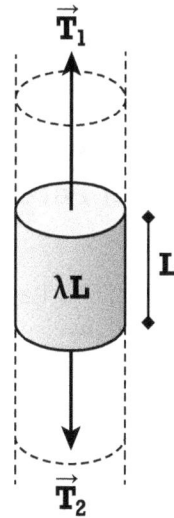

Figure 5.4. A Free-Body Diagram for a small length L of the rope in Figure 5.2. The rope has a mass per unit length of λ (lambda) kg/m. An infinitesimal length dL consequently has a mass $dm = \lambda dL$.

6. **Solve**.

$$m_1 \, a = T - m_1 \, g$$
$$-m_2 \, a = T - m_2 \, g. \tag{5.11}$$

Subtract:

$$(m_1 + m_2) \, a = (m_2 - m_1) \, g$$
$$a = \frac{(m_2 - m_1)}{(m_1 + m_2)} \, g.$$

Plug back in to get T:

$$T = \frac{2m_1 m_2}{(m_1 + m_2)} \, g. \tag{5.12}$$

7. **Check limiting cases (the sanity check).**

$$m_1 = 0 \quad a = g$$
$$m_2 = 0 \quad a = -g \tag{5.13}$$

and in both cases $T = 0$.

Success: It's self-consistent and the limits are reasonable.

5.3   Newton's Third Law: Equal and Opposite Pairs of Forces

Newton's Third Law follows directly from conservation of momentum. Consider a system of two isolated masses, m_1 and m_2, interacting with each other. Newton's Third Law[6] translates to:

> *The forces two bodies exert on each other always are equal in magnitude and opposite in direction.*

Consider the total momentum of the system, a constant:

$$\vec{p}_T = \vec{p}_1 + \vec{p}_2. \tag{5.14}$$

The time derivative of the total momentum is zero:

$$d\vec{p}_T / dt = d\vec{p}_1 / dt + d\vec{p}_2 / dt = 0$$
$$d\vec{p}_1 / dt = -d\vec{p}_2 / dt \tag{5.15}$$
$$\vec{F}_1 = -\vec{F}_2.$$

---

[6] Actioni contrariam semper et aequalem esse reactionem: sive corporum duorum actiones in se mutuo semper esse aequaleset in partes contrarias dirigi.

5.4   Work: The Energy Transferred by a Force

Work is a formal term, and is defined here as the energy transferred by a force when moving an object.[7] The Work done by a force, $\vec{F}$ moving an object over an infinitesimal distance $d\vec{s}$ we define as

$$dW \equiv \vec{F} \cdot d\vec{s} \tag{5.16}$$

where dW is an infinitesimal amount of Work, and $d\vec{s}$ is the infinitesimal displacement of the object under the influence of $\vec{F}$. A constant force of 1 Newton (1 kg-m/s^2) acting over 1 meter (1 m) transfers 1 Joule (1 kg-(m/s)^2) of energy. For a free body, the energy goes into acceleration. In contrast, slowly lifting 1 kg (approximately a quart of milk) one meter transfers approximately 10 Joules of energy to the gravitational field of the milk-Earth system.[8]

5.4.1   Integrating dW Along a Path: Total Energy Expended

To calculate the Work done (energy transferred) over a finite interval we need to integrate the differential quantity along the path of the motion. We thus need to parameterize the value of $\vec{F} \cdot d\vec{s}$ at each point on the path to explicitly form the scalar integrand, and then integrate from the starting point $s = 0$ to the end point s of the path.

5.4.2   Example: Calculating the Work Done by George Moving a Mass on a Closed Path in the Vertical Plane

Figure 5.5 gives an example. George Atwood is initially holding a mass m. He will move it horizontally a distance ℓ, then vertically h, return it horizontally a distance ℓ, finally lowering the mass vertically to its starting place. The path is thus a rectangle lying in the xy plane. We take the origin to be the starting point.

Figure 5.5 also shows the Free-Body Diagram for the mass. Gravity provides a force $-mg\hat{y}$. George supplies the counteracting force $+mg\hat{y}$ to keep the mass stationary. Note that no Work is done when the mass is not moving.

We will follow the path of the mass and compute the Work on each leg.

(1) Horizontal Leg across from $(x, y) = (0, 0)$ **to** $(\ell, 0)$

To move horizontally George has to provide a counteracting equal force $\vec{F} = +mg\hat{y}$, plus a tiny force $\epsilon\hat{x}$ at $x = 0$ to start the horizontal motion. He applies an

---

[7] It is unfortunate that ill-defined colloquial words have been appropriated by physicists to name precisely-defined mathematical quantities. Work is one such; here we define it in Eq. 5.16, and will henceforth refer to it with a capital W.

[8] Traditionally dW is defined as $dW \equiv -\vec{F} \cdot d\vec{s}$, i.e., with the opposite sign. See Section 5.4.3 for why one should skip the minus sign.

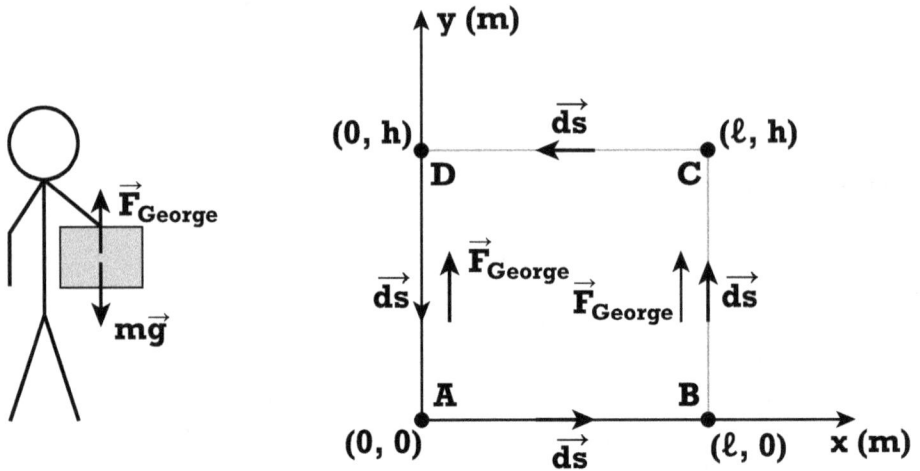

Figure 5.5. The diagram for an example calculation of the Work done by George moving a mass on a closed rectangular path in the vertical plane.

equal tiny force but in the $-\hat{x}$ direction to stop the horizontal motion after the mass has traveled a distance ℓ.[9]

The Work done by George along this leg is found by integrating $dW = d\vec{s} \cdot \vec{F}$ along the length of the path.

$$W_1 = \int_0^\ell (mg\hat{y}) \cdot (dx\hat{x}) = 0 \tag{5.17}$$

which is zero as $\hat{x} \cdot \hat{y} = 0$.

George expended no energy moving the mass perpendicular to gravity.

(2) Vertical Leg up from $(\ell, 0)$ to (ℓ, h)

When the mass is at $(\ell, 0)$, George then adds a tiny short vertical upward force $\epsilon\hat{y}$ to the mass to start the motion, cancelled by an identical tiny short force downwards applied at (ℓ, h) to stop the motion. The Work done by George is:

$$W_2 = \int_0^h (mg\hat{y}) \cdot (dy\hat{y}) = mgh. \tag{5.18}$$

George has transferred energy to the mass-Earth gravitational system.

(3) Horizontal Leg from $(x, y) = (\ell, h)$ to $(0, h)$

Returning along the upper horizontal leg gives zero as for the lower horizontal leg:

$$W_3 = \int_\ell^0 (mg\hat{y}) \cdot (dx\hat{x}) = 0 \tag{5.19}$$

as on the first leg, moving perpendicular to the force, George expends no energy.

---

[9] In addition to being as small as you'd like, these two impulses cancel and so can be completely ignored.

(4) Vertical Leg from $(0, h)$ **to** $(0, 0)$

Returning the mass to the origin downward along the left-hand vertical leg gives back the Work (energy transferred) George did on the upward path.

$$W_4 = \int_h^0 (mg\hat{y}) \cdot (dy\hat{y}) = -mgh. \tag{5.20}$$

On this leg George gains back the energy he lost to the Earth-mass gravitational field raising the mass on the first vertical leg. The total Work done by George along this path is zero:[10]

$$\oint (\vec{F} \cdot \vec{dx}) = 0. \tag{5.21}$$

Note that time does not enter at all in the discussion of Work. George may do this quickly or slowly. Work is energy transferred, and depends only on relative spatial positions and not velocity or acceleration.

5.4.3   Getting the Sign of Work Correct the Feynman Way

After many years of unsuccessfully trying to keep track of the sign of work by diligently keeping track of: (1) the minus sign in the conventional definition; (2) which of the pair of equal-and-opposite forces is F; (3) the limits of integration; (4) the sign of the step dx along the path of integration, and (5) the semantics of Work (*done by* vs. *done on*), I advise not even trying to mathematically track the minus signs, but instead, after calculating the magnitude, ask whether energy was lost or gained. As long as you are consistent in your convention for the energy transfer (i.e., are you keeping track of the energy of George or of the mass), common sense will guide you correctly.

5.4.4   The Recipe for Calculating a Path Integral

To calculate the Work done in Figure 5.5 we did the following steps:

1. Define the path in a coordinate system (draw a labeled diagram);

2. Parameterize the infinitesimal steps $\vec{ds}$ along the path;

3. Form $F \cdot \vec{ds}$ everywhere on the path to get a scalar integrand;

4. Do the (one-dimensional) integral;

5. Check the sign and value for consistency with energy conservation and if the sign comes out wrong, change it.[11]

---

[10] The integral sign with a circle in the middle indicates a closed path of integration.

[11] Anecdotally this rule is Feynman's Last Rule for calculating cross-sections (which are positive-definite) in Quantum Electrodynamics (QED).

Of these, the most difficult concept is the parameterization of the path to form the integral of a scalar quantity over the path. I teach this by the example of Little Red Riding Hood adding flowers to her basket (an integral of a scalar quantity) as she walks to her grandmother's house. It doesn't seem to help. I suggest talking it through with your Study Group, the TAs, and the Instructor. You can also look it up.

5.5   Problem Set 5: Work, Conservative Forces, Simple Path Integrals

Collaborative Learning: The problem sets are designed for collaborative learning, and almost certainly will take too long if you are working alone. You really *must* have a functioning and helpful study group in order to make best use of the course. If your group is not so, ask your Instructor and/or TA to find you a functioning like-minded group that is willing to add another. **However, the work you hand in has to be your own**.

Problem 1: Momentum, Acceleration, and Newton's Second Law

This problem, based on Section 5.1.1, should be the subject of a discussion in your Study Group: it's here only to emphasize an important point.

1. Under what conditions is the more general expression of Newton's Second Law $d\vec{p}/dt = \Sigma_i \vec{F}_i$ not equivalent to the commonly-quoted expression $\vec{F} = m\vec{a}$?

2. Give (at least) two example of physical systems in which the two formulations are not equivalent.

Problem 2: Position, Velocity, and Acceleration in Cartesian Coordinates

1. Draw the xy plane in Cartesian coordinates. Show a position vector $\vec{r}$ that originates (*sic*) at the origin and identifies the position $\vec{r} = (x, y, 0)$ in the $z = 0$ plane.

2. At point $\vec{r}$ draw the unit vectors $\hat{x}$ and $\hat{y}$.

3. Find the expression for the velocity $\dot{\vec{r}} = \vec{v} = d\vec{r}/dt$.

4. Find the expression for the acceleration $\ddot{\vec{r}} = \vec{a} = d\vec{v}/dt$.

Problem 3: Position $\vec{r}$ in Polar Coordinates (See Appendix A, Section A.1.5.)

1. Draw the xy plane in Cartesian coordinates. Show a position vector $\vec{r}$ from origin to the position $\vec{r} = (r, \theta, 0)$ in the $z = 0$ plane. (We use the convention that θ is positive counter-clockwise starting at the positive x-axis.)

2. At point $\vec{r}$ draw the unit vectors $\hat{r}$ and $\hat{\theta}$.

3. Translate your vector to the point $(1,1,0)$ and draw it on your diagram (i.e., keep the components of the vector invariant, but move the point at which it originates to $(1,1,0)$).

4. Express r and θ in terms of x and y. Write a brief note on what you plan to do about the multi-valued nature of θ.[12]

5. Express x and y in terms of r and θ.

Problem 4: Systems—Solving using Free-Body Diagrams—George Atwood and his Famous Machine

George Atwood, not having a good day, tries to lower a barrel from the roof of his house by throwing a rope over a pulley and hanging on one end while standing on the ground. Unknown to George, the barrel weighs 120 kg. George, however, weighs 70 kg. Assuming he and his barrel start at rest, and ignoring the mass of the rope and the pulley, solve for George's subsequent acceleration.

1. Set up the problem by drawing a picture with the appropriate coordinates. I strongly recommend taking the 2 masses to be the parameters M and m rather than the numerical values so that you can check the limiting cases.[13]

2. Draw the Free-Body Diagrams for George and the barrel.

3. Extract the force equations.

4. Write the constraint equations.

5. Verify the number of unknowns is equal to the number of equations.

6. Solve the system of force and constraint equations for the equations of motion of George and the barrel.

7. Plug in the numbers to find George's motion and his fate.

Problem 5: Practice with Path Integrals

1. Consider the vector field $F(\vec{r}, \theta) = r\hat{\theta}$.

   (a) Draw a Cartesian xy coordinate system with origin and labeled axes.

   (b) On your drawing show the magnitude and direction of $\vec{F}$ at three or four points, so that the functional behavior in (r, θ) is clear.

---

[12] The film industry has a mathematically sophisticated solution to the problem of a multi-valued azimuthal angle, which arises when the script needs to uniquely specify camera motions between two scenes. If you are curious about the mathematics, see *Visualizing Quaternions* by A. Hanson [38].

[13] If you plug in the numbers before solving for the form of the answer you have no way of checking that your solution is correct.

(c) Consider a closed path S consisting of a circle of radius R centered on the origin. Calculate the integral:

$$\oint_S (\vec{F} \cdot \vec{ds}) \tag{5.22}$$

where $\vec{ds}$ is an infinitesimal step along the path.[14]

2. Follow the same steps (a, b, and c) to calculate the path integral

$$\oint_S (\vec{F} \cdot \vec{ds}) \tag{5.23}$$

for the vector field $F(\vec{r}, \theta) = r\hat{r}$.

Problem 6: Work Done on a Closed Path: Conservative and Non-Conservative Forces

1. A Conservative Force
   Consider the gravitational force exerted by mass M on mass m:

$$\vec{F}(r) = -\frac{GmM}{r^2}\hat{r} \tag{5.24}$$

where $\vec{r} = r\hat{r}$ is the vector from M to m, and the minus sign indicates the force is attractive (unlike the Coulomb force, which comes in 2 flavors, gravity is only attractive).

Calculate the work done by gravity on a planet moving around a star in a circular orbit of radius R. (Please set up the integral and calculate the integrand.)

2. A Non-Conservative Force: Friction
   A 100-kg weight is moved around a circular path of radius 1m on a table with a coefficient of sliding friction[15] of $\mu_k = 0.7$.

   (a) Draw a diagram with x-y axes and the circular path characterized by the polar coordinates r and θ. Show a point on the path and identify the corresponding coordinates x, y and r, θ on the diagram.

   (b) Draw a Free-Body Diagram of the weight showing the direction and magnitude of all forces on it. On the diagram show $\vec{F}$ and $\vec{ds}$ at several points along the path.

   (c) Calculate the work done by an explicit integration over the path. Please work in polar coordinates (Appendix A, Section A.1.5).

---

[14] Please do not be apprehensive—this is only a single-variable integral—the scalar product of two vectors gives you a scalar function that you know how to integrate. Parameterize the integral by writing a little step $\vec{ds}$ along the circular path in terms of R, θ, $\hat{r}$, and $\hat{\theta}$, and taking the dot product with $\vec{F}$ to produce an integrable function.

[15] Friction should have been covered in your AP Physics class. If not, please look it up.

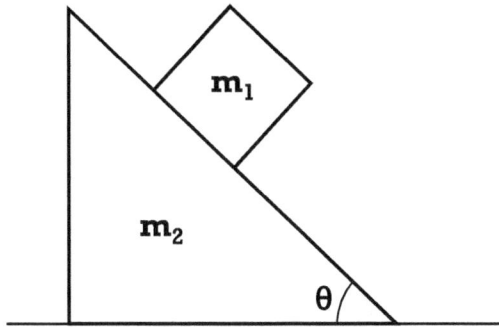

Figure 5.6. A block on a friction-less wedge (image from the web).

Problem 7: A Non-Conservative Force

Consider a 2-dimensional vector field that describes a force $\vec{F} = A_0(x\hat{y} - y\hat{x})$.

(a) Using the line-integral definition of a conservative force, $\oint \vec{F} \cdot d\vec{s} = 0$, show that this force is not conservative.

(b) Write this force in polar coordinates.

Challenge Problem 8: A Free-Body Diagram for a Block Sliding on a Wedge; Sliding on a Surface (More difficult.)

Consider a block of dry ice sitting on a teflon wedge on a flat surface, as shown in Figure 5.6. Taking the friction between the block and the wedge, and the wedge and the surface, to be zero, find the subsequent accelerations of the block and the wedge after they are released.

(a) Define a coordinate system and the coordinates of the wedge and of the block;

(b) Write the two equations (x and y, e.g.) of motion for the block;

(c) Write the two equations of motion for the wedge;

(d) Write the constraint equations for the block;[16]

(e) Solve for the accelerations of the block;

(f) Solve for the accelerations of the wedge;

(g) Explicitly check the limiting cases of $\theta = 0$ and $\theta = 90$ degrees

(h) **Double Challenge** (ahead of ourselves—Chapter 6, but easy): Use conservation of energy and momentum to find the velocities of the block and the

---

[16] If we parameterize the top surface of the wedge by a line slanted up and to the left from the lower tip of the wedge, the constraint is that the block is sitting on the line. Note this does not violate our injunction of not doing kinematics in a non-inertial frame; the constraint equation is purely geometrical, relating the position of the block to the wedge. No accelerations and forces are involved.

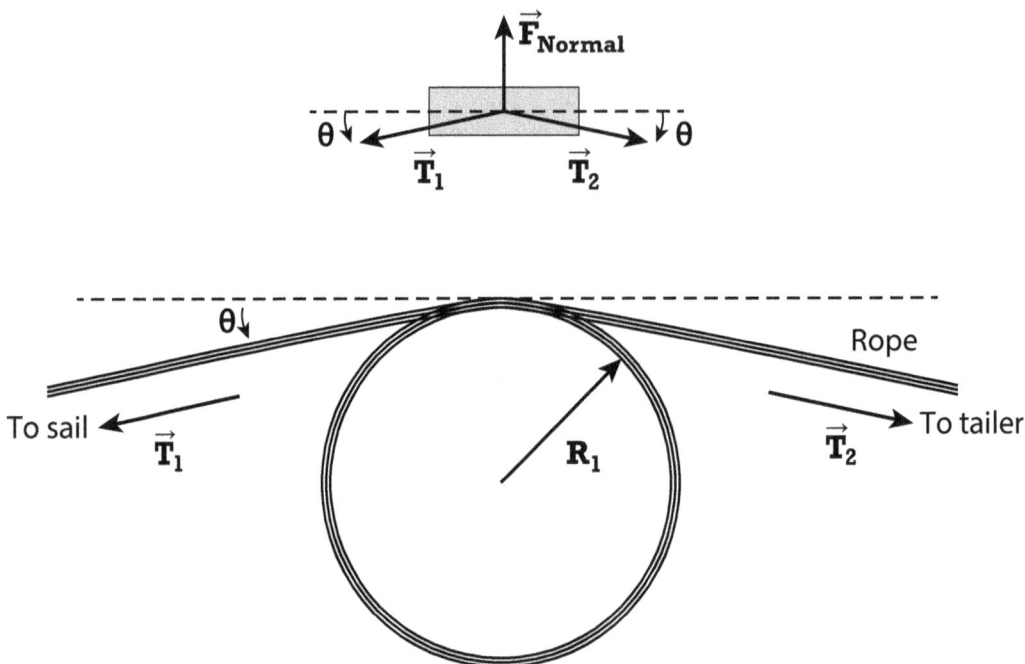

Figure 5.7. A small piece of line (rope) on the capstan of a sailboat (Problem 9).

wedge after the block has arrived at the flat surface and the two have separated. (Conservation laws are wonderful.)

Challenge Problem 9: (More Difficult) Free-Body Diagrams, Friction

This problem requires a careful (large, clear) diagram showing a small piece of line (rope) on the capstan, a thoughtful assignment of parameters, and a good Free-Body Diagram. Like many of these problems, once you know how to set it up it isn't so hard. Note that the tension in the line isn't constant due to the friction on the capstan. Discussion with your group should clarify the forces in the line.

On a sailboat the tension of a sheet (a sheet is a line connected to a sail) that transmits the force of a sail may be taken up by a winch, which consists of a cylindrical capstan around which the line is wrapped several turns. A member of the crew (called a tailer), puts a tension on the end of the line exiting the winch. Using a Free-Body Diagram, show that the force with which the tailer has to pull on the tail of the line (T_2) is related to the pull of the sail (T_1) by:

$$T_2 = T_1 e^{-\mu\theta} \tag{5.25}$$

where μ is the coefficient of friction of the line on the capstan, and θ is the total angular extent of the line around the capstan (e.g., one turn is 2π, 2 turns is 4π, etc.).

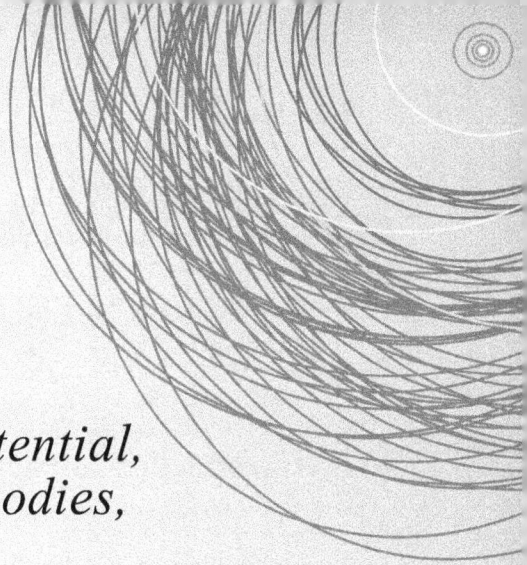

PART II

Dynamics: Gravitational Potential, Angular Momentum, Rigid Bodies, and Central Force Motion

Our Moon, our sister planet Jupiter, Jupiter's moons Europa, Ganymede and Callisto, seen over the Wasatch Mountains. The first measurement of the finite speed of light was in 1676 by the Danish astronomer and mathematician Ole Rømer, who used the regular transit of one of Jupiter's moons as ticks of a "clock" to get a value within 28% of the modern value.

The moons of Jupiter, several of which are shown [41], have played a remarkable role in the story of the perception of light as being instantaneous (Galilean) versus finite (relativistic). Ole Rømer, a Danish astronomer/mathematician measured the speed of light at the Paris Observatory in 1676, using a method proposed by Galileo to determine terrestrial longitude for navigation, and a telescope, itself adopted and improved by Galileo. The transit of the planets across the face of Jupiter, provides a stable precision clock.[1] However, the distance between Jupiter and the Earth varies by about 20 minutes as the Earth goes around its orbit. Rømer had been working with Cassini (of Saturn's rings) in Paris, and went on to measure the phase shift; his value for c was 2.2×10^8 m/s, within 30% of the current value [25]. Remarkable and lovely.

Introduction to Part II

In Part II we extend the non-relativistic kinematics introduced in Part I to the topics in dynamics covered in a first-year one-quarter Honors Course. The development in Part I of a more sophisticated mathematical language than is typical allows us to go somewhat deeper and faster in several topics. The level of presentation and the difficulty of the problems are intended to be no different from in Part I, tractable to motivated Honors high school or college students working collaboratively with a like-minded group and strong instructional support. Part II may look more difficult mathematically (e.g., deriving acceleration in polar coordinates, or the equation for elliptical orbits), but I believe the real difficulties lie in the disconnect between our developed intuition and actual behaviors, e.g., the motions of spinning rigid bodies or planets in elliptical orbits.[2] These are topics that are "open-ended." One doesn't just learn them and be done as in a typical high school physics course. Instead we can use new language to observe more deeply. The learning environment consequently must continue to be collaborative; working with your study group is essential. Strong support from TAs and the instructor, active suppression of competitive urges (the goal is that all students do well), and high-quality solutions to the problem sets are necessary. The hope is that the course opens up new curiosities and more powerful tools to explore them for a broader demography of students than typical in the traditional Honors course.

Paul Adrien Maurice Dirac was a key architect in Quantum Mechanics and the electromagnetic theory of particles.[3] Sometime in the late 1970s Dirac gave a public lecture in the Ramsey Auditorium at Fermilab. Dirac was old and very frail, but still himself. The auditorium was packed, with physics luminaries in the front row. At the end of the lecture, one of the luminaries (I forget who now) asked: "Paul, how could you have proposed the proton as the anti-electron?" (The proton is 2000 times heavier than the electron, and the masses of the e^- and e^+ should be the same.) Dirac replied,

---

[1] It is deeply pleasing to note the similarity to Einstein's example of Casals' clock, and the commonality of invoking the speed of light three centuries apart. The difference, however, is that Casals' measurement depends on the invariance of the speed of light in two frames, while Rømer's measurement depended only on the difference in path length across the Earth's orbit in basically a single frame of reference.

[2] And bad habits cultivated and reinforced by AP Physics should not be ignored.

[3] See *The Strangest Man*, by Graham Farmelo, in Recommended Reading.

"I did not think it correct, but I was young, and in those days one did not go around inventing particles." Someone from the front row—I think it was Weisskopf—said, "Einstein had suggested the existence of the positron." Dirac looked very elderly and terribly unhappy—I remember thinking he might actually cry—but then he totally brightened up, pointed one finger in the air, and said, "Ah, but he was only guessing!"

The following is from a lecture Dirac delivered on the presentation of the James Scott Prize in 1939: Proceedings of the Royal Society (Edinburgh) Vol.59, 103839.

The Relation Between Mathematics and Physics

The trend of mathematics and physics towards unification provides the physicist with a powerful new method of research into the foundations of his subject, a method which has not yet been applied successfully, but which I feel confident will prove its value in the future. The method is to begin by choosing that branch of mathematics which one thinks will form the basis of the new theory. One should be influenced very much in this choice by considerations of mathematical beauty. It would probably be a good thing also to give a preference to those branches of mathematics that have an interesting group of transformations underlying them, since transformations play an important role in modern physical theory, both relativity and quantum theory seeming to show that transformations are of more fundamental importance than equations.[4] Having decided on the branch of mathematics, one should proceed to develop it along suitable lines, at the same time looking for that way in which it appears to lend itself naturally to physical interpretation.

---

[4] This is a remarkable statement; it would be interesting to explore the relation to Emmy Noether's work in Göttingen and elsewhere (See Ref. [13] and *The End of the Certain World* in Recommended Reading.)

CHAPTER 6

Conservative Forces, the Potential $V(r)$, and Force $\vec{F} = -\vec{\nabla}V(r)$

6.1 Force as the Derivative of a Potential

Consider a ball on the inside wall of a cylindrically symmetric container as shown in Figure 6.1. To solve the motion using forces as we did in Chapter 5, we would draw a Free-Body Diagram of the ball showing the normal force exerted by the wall and the force of gravity. The normal force is a function of the shape of the container wall at the position of the ball; gravity is everywhere downward. Breaking the problem into components we could solve for the component of acceleration parallel to the wall at the position of the ball.

However, there is a powerful abstraction that allows a much easier solution. The shape of the wall determines the height of the ball above some zero point, which for convenience we can take as the lowest point of the container.[1] We can define a scalar function, the Potential $V(y)$, which in this case depends only on the height y.

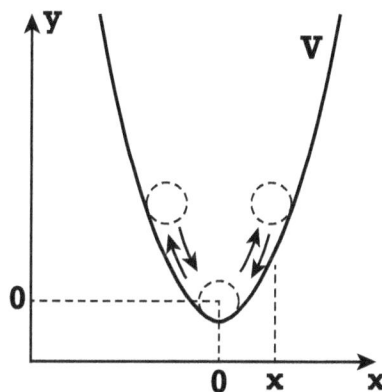

Figure 6.1. A potential well. By definition the first derivative of the potential is zero at the minimum (if it were not, one could go lower). The second derivative gives a restoring force (restoring means back towards the equilibrium position) linearly proportional to the displacement.

The force accelerating the ball will be proportional to the derivative of $V(y)$ with respect to y. This derivative is easier to calculate than the integrals needed to start from the components of F.

The abstraction from force, a vector, to potential energy, a scalar, is in fact natural to us, although we may not have thought about it. When riding a bike, we

---

[1] This is arbitrary: only *differences* in height matter.

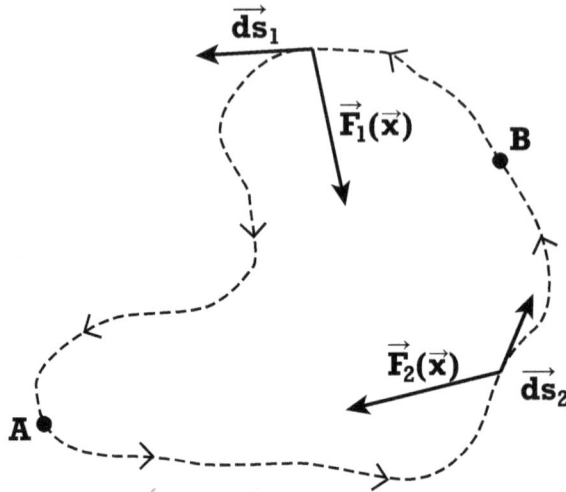

Figure 6.2. The Work done $\int_A^B \vec{F} \cdot \vec{ds}$ along paths 1 and 2.

are aware of going uphill or downhill, i.e., changes in potential energy, rather than being conscious of the normal force from the road and the vertical force of gravity on the bike. Peddling uphill requires more energy than coasting downhill, and the steepest way down is fastest. All this is deeply intuitive.

In this chapter we will develop the mathematics to express our intuitive understanding concisely with force being the negative gradient (i.e., downhill) of the potential energy. Forces for which we can define a potential energy are called conservative forces.

6.2   Conservative Forces

A force is defined to be conservative if the Work done along a path connecting two points, A and B, depends only on the spatial coordinates of A and B and not on the path.

6.2.1   The Integral Condition $\oint \vec{F} \cdot \vec{ds} = 0$ for a Conservative Force

Consider the two paths shown in Figure 6.2. If the force $\vec{F}$ is conservative, the work along path 1 is equal to the work done in path 2.

$$\int_A^B \vec{F} \cdot \vec{ds}_1 = \int_A^B \vec{F} \cdot \vec{ds}_2. \tag{6.1}$$

Subtracting the right-hand side from both sides:

$$\int_A^B \vec{F} \cdot \vec{ds}_1 - \int_A^B \vec{F} \cdot \vec{ds}_2 = 0 \tag{6.2}$$

where $\vec{ds}_1$ and $\vec{ds}_2$ are differential elements of two different paths connecting A and B.

Now consider returning from B to A on path 2:

$$\int_B^A \vec{F} \cdot \vec{ds}_2 = -\int_A^B \vec{F} \cdot \vec{ds}_2. \tag{6.3}$$

Substituting into Equ. 6.2:

$$\int_A^B \vec{F} \cdot \vec{ds}_1 + \int_B^A \vec{F} \cdot \vec{ds}_2 = 0. \tag{6.4}$$

The combined paths of the two integrals corresponds to going from A to B on path 1 and returning on path 2, which is a closed path. We have proved that the integral over the closed path from A to B and back again is identically zero:

$$\oint \vec{F} \cdot \vec{ds} = 0. \tag{6.5}$$

The integral of a conservative force over any closed path is zero.

This is the integral definition of a conservative force. Any conservative force must satisfy the condition; any force that satisfies it is conservative.

6.2.2   Examples of Conservative and Non-Conservative Forces

The Earth's gravitational field and an electric field from fixed electric charges are both examples of conservative forces. The force at any point along the path depends only on spatial coordinates, and the Work done along any closed path will be zero.

In contrast, friction is a non-conservative force. The Work done to slide a block in a path on a table will be approximately proportional to the length of the path, as the force $\vec{F}$ is opposite to the direction of motion $\vec{ds}$ at every point on the path and roughly independent of velocity.[2] The force required for constant velocity $\vec{F}$ (i.e., to oppose the frictional retarding force) is proportional to the normal force, in this case the weight of the block, with a coefficient of friction μ. Friction is dissipative, i.e., energy goes into heating[3] the block and the table.

6.3   Potential

Let us define one point in space, call it $\vec{r}_0$, from which we can integrate $dV = -\vec{F} \cdot \vec{dx}$ to a point $\vec{r}$ along any path, where we have substituted the potential $V(\vec{r})$ for W.

---

[2] As long as the velocity is non-zero. Once the block is stopped, it takes extra force to break additional atomic/molecular bonds to get the block moving again.

[3] Colloquially "is lost;" but of course not lost, just unaccounted for in our bookkeeping of macroscopic kinetic and potential energy.

Unlike Work, $V(\vec{r})$ represents a scalar function over all space that has a value at each point $\vec{r}$ equal to the energy difference $V(\vec{r}) - V(\vec{r_0})$. As only the differences in energy at different locations determine the force, we are free to set the reference potential $V(\vec{r_0})$ to an arbitrary value. Defining $V(\vec{r_0}) = 0$, the potential is simply $V(\vec{r})$.

6.3.1   Force as the Gradient of the Potential

To find the force given the potential:

$$dV = -\vec{F} \cdot \vec{dx}$$
$$dV = -(F_x dx + F_y dy + F_z dz). \tag{6.6}$$

Writing out the components explicitly:

$$F_x = -\frac{\partial}{\partial x} V(x, y, z)$$

$$F_y = -\frac{\partial}{\partial y} V(x, y, z) \tag{6.7}$$

$$F_z = -\frac{\partial}{\partial z} V(x, y, z)$$

where the symbol $\frac{\partial}{\partial x} V(x, y, z)$ stands for the partial derivative[4] of the potential with respect to x.

6.4   The Differential Operator ∇ and the Gradient $\nabla f(x, y, z)$

We define the vector differential operator ∇(Del):

$$\nabla = \hat{x} \frac{\partial}{\partial x} + \hat{y} \frac{\partial}{\partial y} + \hat{z} \frac{\partial}{\partial z}. \tag{6.8}$$

We can write the force as the negative *gradient* of the potential:

$$\vec{F} = -\nabla V(x, y, z)$$
$$\vec{F} = -(\hat{x} \frac{\partial V(x, y, z)}{\partial x} + \hat{y} \frac{\partial V(x, y, z)}{\partial y} + \hat{z} \frac{\partial V(x, y, z)}{\partial z}). \tag{6.9}$$

---

[4] In taking a partial derivative with respect to x, the other two parameters, y and z, are held constant; see Appendix A, Section A.1.2.

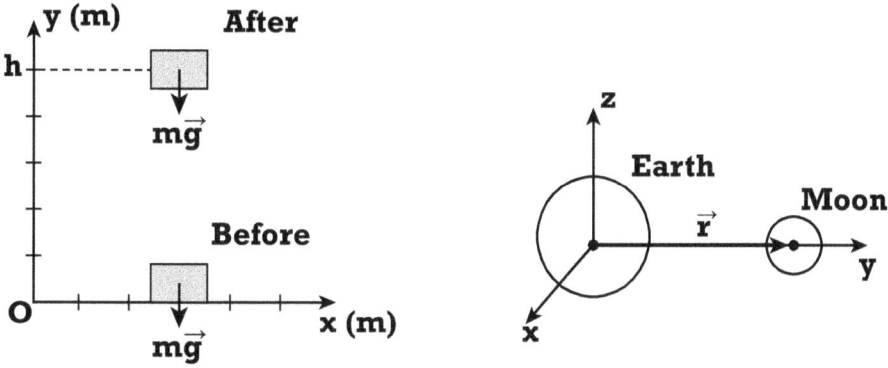

Figure 6.3. Left: A mass is lifted a height h. Right: A moon is orbiting a planet at radius r.

Note that in Eq. 6.9 the gradient operates on a scalar function and returns a vector.

Also note the (important) minus sign.[5] By definition, the gradient of a scalar function is a vector pointing on the *steepest* path *uphill*. The force $\vec{F}$ on an object at that point corresponds to a vector pointing on the steepest path *downhill*, and hence the minus sign in $\vec{F} = -\nabla V$.

6.4.1 Calculating Force from the Potential: Two Gravitational Examples

The force at any point in space is the negative gradient of the potential, evaluated at that point. We show two simple examples in Figure 6.3, gravity on the Earth's surface and gravity in a moon-planet system.

6.4.1.1 Example I: Gravity on the Earth's Surface

The left-hand panel of Fig. 6.3 shows a mass m at a height h above the floor. The potential is $V(\vec{r}) = mgy$. Calculating the force $\vec{F}$ from the potential V:

$$\vec{F} = -\nabla V$$
$$= -(\hat{x}\frac{\partial}{\partial x} + \hat{y}\frac{\partial}{\partial y} + \hat{z}\frac{\partial}{\partial z}) \, mgy$$
$$= -(\hat{x}\frac{\partial}{\partial x}mgy + \hat{y}\frac{\partial}{\partial y}mgy + \hat{z}\frac{\partial}{\partial z}mgy) \qquad (6.10)$$
$$= 0\,\hat{x} - mg\,\hat{y} + 0\,\hat{z}$$
$$\vec{F} = -mg\,\hat{y}.$$

---

[5] Unlike in the definition of Work, this minus sign matters. The gradient points uphill; one skis downhill.

6.4.1.2   *Example II: A Moon-Planet System*

The right-hand panel of Fig 6.3 shows a moon of mass m orbiting a planet of mass M. The potential is

$$V(\vec{r}) = -\frac{k}{r}$$

$$= -\frac{k}{(x^2 + y^2 + z^2)^{1/2}} \qquad (6.11)$$

$$= -k\,(x^2 + y^2 + z^2)^{-1/2}$$

where $k = GmM$ is a constant. Calculating the x-component of the force $\vec{F}$ from the potential V in Eq. 6.11:

$$\vec{F} = -\nabla V$$

$$F_x = -\frac{\partial}{\partial x}(-k\,(x^2 + y^2 + z^2)^{-1/2}\,)$$

$$= +k\,(-\frac{1}{2})\,((x^2 + y^2 + z^2)^{-3/2}\,(2x)) \qquad (6.12)$$

$$= -kx\,(x^2 + y^2 + z^2)^{-3/2}$$

$$F_x = -k\,(\frac{x}{r^3}).$$

By symmetry the y and z components will have the same form. Putting the y and z components back into the vector $\vec{r}$ and using the identity $\vec{r} = r\hat{r}$, where $\hat{r}$ is the unit vector in the $\vec{r}$ direction:

$$\vec{F} = -k\,\frac{x\hat{x} + y\hat{y} + z\hat{z}}{r^3}$$

$$= -k\,\frac{\vec{r}}{r^3}$$

$$= -k\,\frac{r\hat{r}}{r^3} \qquad (6.13)$$

$$\vec{F} = -\frac{k}{r^2}\,\hat{r}$$

i.e., $\vec{F}$ is an inverse-square attractive force.

6.4.2   The Differential Condition $\nabla \times \vec{F} = 0$ for a Conservative Force

In Section 6.4 we introduced the differential *vector* operator Del, $\nabla \equiv \hat{x}\frac{\partial}{\partial x} + \hat{y}\frac{\partial}{\partial y} + \hat{z}\frac{\partial}{\partial z}$, and applied it to a *scalar* function to find the *vector* representing the maximum slope:

$$\vec{F} = -\vec{\nabla} V(x,y,z). \tag{6.14}$$

Note that the gradient is a vector operator acting on a scalar function, returning a vector.

The vector product (cross product) of Del and a vector $\vec{A}$, $\nabla \times \vec{A}$, is called the "curl" of $\vec{A}$ (see Appendix A, Section A.1.7). At each point of a vector field F the magnitude of curl represents the limit of a closed-loop integral, $\oint \vec{F} \cdot \vec{ds}$, as the area goes to zero.[6] A field with non-zero curl will have non-zero integral over a closed path and vice versa. Note that the curl is a vector operator acting on a vector, returning a vector.

The differential form of the condition that a force be conservative is:

$$\nabla \times \vec{F} = 0. \tag{6.15}$$

Writing out the ith component of Eq. 6.15 in index notation:

$$(\nabla \times \vec{F})_i = \nabla_j F_k - \nabla_k F_j \tag{6.16}$$

where the indices i, j, k run from 1 to 3 in cyclic order.[7] For example, when $i = 1$,

$$(\nabla \times \vec{F})_1 = \nabla_2 F_3 - \nabla_3 F_2. \tag{6.17}$$

When $i = 2$

$$(\nabla \times \vec{F})_2 = \nabla_3 F_1 - \nabla_1 F_3. \tag{6.18}$$

When $i = 3$

$$(\nabla \times \vec{F})_3 = \nabla_1 F_2 - \nabla_2 F_1. \tag{6.19}$$

Translating back from the more general index notation (i,j,k) for the 3 component directions to the notation using (x,y,z), the components of $\vec{F}$ are:

$$(\nabla \times \vec{F})_x = \nabla_y F_z - \nabla_z F_y$$
$$(\nabla \times \vec{F})_y = \nabla_z F_x - \nabla_x F_z \tag{6.20}$$
$$(\nabla \times \vec{F})_z = \nabla_x F_y - \nabla_y F_x.$$

What we have gained with the introduction of the curl is that testing whether a force $F(x,y,z)$ is conservative is *much* easier by calculating the derivatives $\nabla \times \vec{F}(x,y,z)$ than by doing path integrals around loops.

---

[6] This is Stokes' Theorem. The proof is sweet and not hard, but is of lower priority for a one-quarter (read time-pressed) introductory course in Classical Mechanics. I spend the appropriate time on it in the second quarter Honors course on Electricity and Magnetism. You can look it up, it's neat and important.

[7] Cyclic order is any 3 consecutive numbers in the sequence 123123123 ad infinitum, i.e., 123, 231, or 312. Anticyclic occurs when 2 integers in a cyclic triplet are switched: i.e., 132, 213, or 321.

6.4.3   Conservation of Energy: E=V+T

Consider a conservative force $\vec{F}$ applied along the positive x axis to a mass m. The acceleration of the mass is described by Newton's Second Law:

$$ma_x = F_x. \tag{6.21}$$

So as not to cancel infinitesimals,[8] we will calculate with small finite quantities and take the limit after some algebraic manipulation:

$$
\begin{aligned}
a &= \frac{dv}{dt} \\
&= \frac{dv}{dx}\frac{dx}{dt} \\
&= \frac{dv}{dx}v \\
&= v\frac{dv}{dx}.
\end{aligned} \tag{6.22}
$$

The Work *done* (energy transferred) pushing m from 1 to 2 is $dW = F_x dx$. The energy gained by m going from point 1 to point 2 is then:

$$
\begin{aligned}
\int_1^2 F dx &= \int_1^2 ma\,dx \\
&= \int_1^2 mv(\frac{dv}{dx})dx \\
&= \int_1^2 m(v\,dv).
\end{aligned} \tag{6.23}
$$

The integral is straight-forward, as $v\,dv$ is a "perfect differential," i.e., $\int_1^2 v\,dv = \frac{1}{2}v^2$. Putting it all together:

$$\int_1^2 F dx = \frac{1}{2}mv^2\Big|_1^2 = \frac{1}{2}mv_2^2 - \frac{1}{2}mv_1^2. \tag{6.24}$$

The right-hand side $\frac{1}{2}mv_2^2 - \frac{1}{2}mv_1^2$ is the change in kinetic energy ΔT; the left-hand side is the work done by the force, with energy supplied from the potential, ΔV. We have derived that the total energy is conserved, $E_{after} = E_{before}$.

$$
\begin{aligned}
\Delta T &= -\Delta V \\
\Delta(T + V) &= \Delta E = 0.
\end{aligned} \tag{6.25}
$$

---

[8] This is to placate the math majors; it really doesn't matter, and the world is probably lumpy at very small distances anyhow.

6.5   Our World of Macroscopic Conservative Forces: Gravity and Electro-Magnetism

Many of the phenomena of our daily lives are described by two conservative macroscopic forces, gravitation and the electrostatic force. We briefly state the Laws.

6.5.1   The Gravitational Force: Newton's Law of Gravitation

The gravitational force between two bodies of mass m and M is given by:

$$F_G = \frac{GmM}{r^2} \tag{6.26}$$

where G is Newton's Constant, $G \approx 7 \times 10^{-11}$ N-m^2/kg^2, where N is the unit of force, the Newton. A force of 1 Newton accelerates a mass of 1 kg by an acceleration of 1 m/sec^2. Note that two 1 kg masses a meter apart attract each other with a force of approximately seventy pico-Newtons (pico is 10^{-12}).

6.5.2   The Electrostatic Force: Coulomb's Law

The electrostatic (Coulomb) force between two charges q_1 and q_2 is given by:

$$F_C = k\frac{q_1 q_2}{r^2} \tag{6.27}$$

where k is the Coulomb Constant, $G \approx 9 \times 10^9$ N-m^2/C^2, and C is the SI unit of charge, the Coulomb. Charge is quantized in units of e; the charge of a proton or electron has magnitude $e = 1.6 \times 10^{-19}$ C. Note that while two like-sign charges of magnitude 1 Coulomb each[9] a meter apart repel each other with a force of 9×10^9 (9 billion) Newtons, two electrons a meter apart repel each other with a force of approximately 2×10^{-28} Newtons.

6.6   Hooke's Law and Simple Harmonic Motion

Hooke's Law[10] states that for "elastic" materials, small displacements from an equilibrium condition impart a *linear* restoring force. Elastic forces are conservative, and many day-to-day interactions involve materials that are approximately elastic. Elasticity is due to the details of underlying atomic and molecular structures, which

---

[9] Good luck with that.

[10] Robert Hooke (1635-1703) was born on the Isle of Wight. Hooke's statement of the Law is wonderfully succinct: *ut tensio, sic vis*; (*as extension, thus force*, i.e., the force is linearly proportional to the displacement from equilibrium).

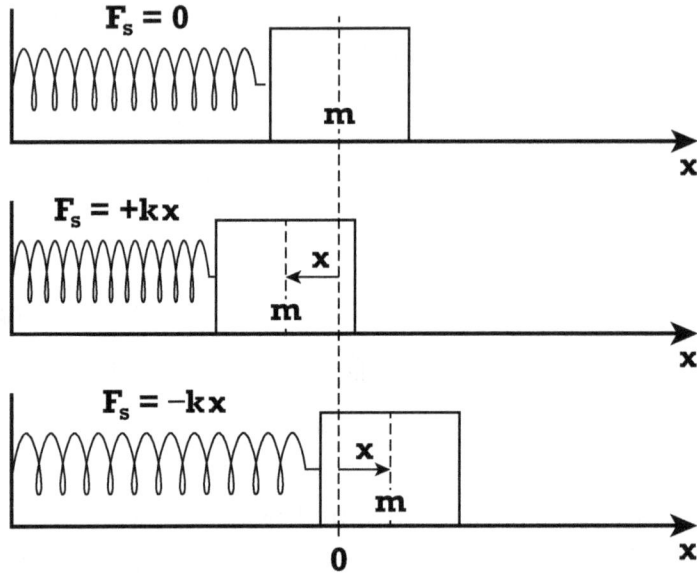

Figure 6.4. A block connected to a spring sliding back and forth on a slippery table. The block undergoes Simple Harmonic Motion, i.e., the position, the velocity, and the acceleration all are described by sines and cosines with fixed relative phases.

are themselves beyond the scope of this (macroscopic) course. We will see that Hooke's Law corresponds to the first non-zero term in the Taylor expansion of the potential around the minimum (the equilibrium position).

6.6.1   The Linear Restoring Force and the "Spring Constant" k

The coefficient k of the linear restoring force is often called the spring constant.

6.6.2   Harmonic Functions: Sines and Cosines as Solutions

Consider a block sitting on a table, connected to a fixed vertical surface by a spring, as shown in Figure 6.4. We will ignore friction.

The equation of motion is:

$$m\ddot{x} = -kx$$

$$m\ddot{x} + kx = 0 \tag{6.28}$$

with the solution:
$$x(t) = A \sin \omega t + B \cos \omega t$$

where $\omega = \sqrt{\frac{k}{m}}$ is the angular frequency,[11] and A and B are constants to be set by the initial conditions found by setting $t = 0$ for both the position x and the velocity $\dot{x}$ (2 unknowns and 2 constraints).

---

[11] See Appendix A, Section A.1.4. Remember the frequency of oscillation is $\nu = \omega/2\pi$, where ω is the angular frequency.

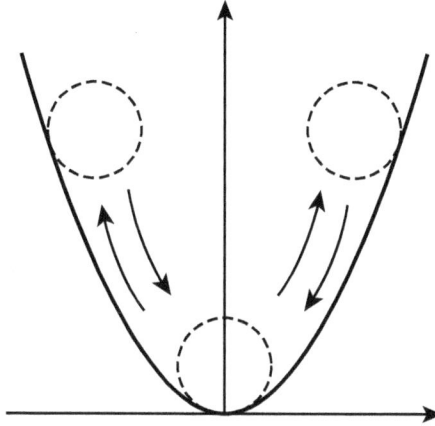

Figure 6.5. An enlargement of the minimum of the potential well. By definition the first derivative of the potential is zero at the minimum (if it were not, one could go lower). The second derivative gives a restoring force linear in the displacement.

Taking the time derivatives to get the velocity and acceleration:

$$x(t) = A \sin \omega t + B \cos \omega t$$
$$\dot{x}(t) = \omega A \cos \omega t - \omega B \sin \omega t \qquad (6.29)$$
$$\ddot{x}(t) = -\omega^2 A \sin \omega t - \omega^2 B \cos \omega t.$$

6.6.3   Taylor Expansion of the Bottom of a Potential Well: Simple Harmonic Oscillations

Simple harmonic motion is not limited to the case of a pure quadratic potential such as (by definition) that for a simple spring. Consider a non-quadratic potential well corresponding to a more complex physical process,[12] as shown in Figure 6.5. Expanding the potential around the minimum value in a Taylor expansion (See Appendix A, Section A.1.3):

$$V(x) = V(0) + \frac{1}{1!}\frac{\partial V}{\partial x}\bigg|_{x=0} x + \frac{1}{2!}\frac{\partial^2 V}{\partial x^2}\bigg|_{x=0} x^2 + \ldots \frac{1}{n!}\frac{\partial^n V}{\partial x^n}\bigg|_{x=0} x^n + \ldots$$

$$V(x) - V(0) = \frac{1}{2}kx^2 + \ldots.$$

$$(6.30)$$

The value of the potential at the minimum, $V(0)$, is arbitrary, doesn't enter into the motion, and can be set to zero. The first derivative is zero at the minimum by

---

[12] A pendulum, for example, has a more complex potential than purely quadratic, being proportional to the cosine of the displacement angle. See Problem 5 at the end of this chapter.

definition. Consequently the second derivative will determine the potential, and hence the motion, in the limit of small oscillations:

$$V(x) = \frac{1}{2}kx^2. \tag{6.31}$$

The spring constant k is given by $k \equiv \frac{\partial^2 V}{\partial x^2}\big|_{x=0}$. For small oscillations around the minimum we can ignore the higher order terms and the motion will be approximately harmonic with frequency $\omega = \sqrt{k/m}$, i.e., determined by the second derivative of the potential at the equilibrium point.

6.7   A Taking Stock; A Partial Summary

In Chapters 4 and 5 of Part I we made the transition to the non-relativistic limit in which the speed of light is infinite, introducing Newton's brilliant Second and Third Laws. We then introduced Work, the transfer of energy by the exertion of a force over a distance, and calculations of Work using simple path integrals.

In this chapter we used these tools to define conservative forces for which we can define a *scalar* function, the potential $V(\vec{r})$, where the *vector* force $\vec{F}$ is the negative gradient of the potential. Encoding the dynamics of motion in 3-dimensions in a scalar (1-dimensional) function from which one gets the vector behavior by taking derivatives is computationally much easier and more robust than keeping track of forces and motions in 3-space. The transition from physical 3-vectors to more abstract scalars is a step along the path to a more sophisticated classical mechanics, and mathematically if not intuitively, quantum mechanics.

The expansion of an arbitrary potential well[13] around its minimum in a Taylor expansion (Appendix A, Section A.1.3) leads naturally to a quadratic potential as the first non-trivial non-zero term. A quadratic potential returns a linear restoring force so that a particle trapped in a 1-dimensional well oscillates across the minimum. This is Simple Harmonic Motion (SHM), and is ubiquitous in mechanical systems and our daily lives.[14]

Harmonic Functions, the solutions to SHM, have a wealth of applications and interest. The solutions, which are of the form $A \cos nx$ and $B \sin nx$, form a complete orthonormal set of "basis vectors" that "span the space" of differentiable functions on the interval $(-1, 1)$ [26]. Modern communications, electrical engineering, medical and other imaging, for example, depend on analyses in frequency space.[15]

---

[13] A well is a region surrounding a minimum in the potential.

[14] For an excellent more-realistic treatment including damping and driving forces, see the text by Marion in *Recommended Reading*, for example.

[15] In addition to the analytic importance, there is a natural importance: all vision, hearing, touch—all the senses—involve electromagnetic signals, and consequently frequency. And of course, Music and the Arts: it is no accident that the sine and cosine solutions to SHM are called "harmonic functions."

6.8   Problem Set 6: Conservative Forces, Potential $V(\vec{r})$, $\vec{F} = -\nabla V(\vec{r})$, SHM

Problems with answers, and recycled problems: There are a limited number of beginning mechanics problems, and so one can find answers to most by searching on the web. We trust you not to copy.

Units and Constants: Please use MKS (SI) units where appropriate.

The Newton: The SI unit of force is the Newton, N. A force of 1 N accelerates a mass of 1 kg by 1 m/s^2. Its units are necessarily that of mass times acceleration, i.e., kg-m-s^{-2}.

Newton's Constant: $G \equiv 6.7 \times 10^{-11}$ Nm^2/kg^2. Two masses of 1 kg, 1 meter apart, exert a gravitational force on each other of $F = \frac{Gm_1 m_2}{r^2} = G$ Newtons, i.e., 67 pico-Newtons.

Please work in Cartesian coordinates. Be careful of the sign of F.

Problem 1: Finding $\vec{F}$ from $V(x)$

1. **The 1-dimensional SHM Potential $V(x) = kx^2$**
   Consider the potential $V(x) = kx^2$ where k is a constant.
   Find the force as a function of x.

2. **The 1-dimensional Potential $V(x) = kx^n$**
   Consider the potential $V(x) = kx^n$ where k is a constant and n is an integer.
   Find the force as a function of x.

3. **The Force for a 1-Dimensional Exponential Potential**
   Consider the potential $V(x) = e^{k|x|}$ where k is a constant.
   Find the force as a function of x.

Problem 2: Finding $\vec{F}$ from $V(\vec{r})$

1. **The 3-dimensional Potential $V(\vec{r}) = kr^2$**
   Consider the potential $V(\vec{r}) = kr^2$ where k is a constant and $r = \sqrt{x^2 + y^2 + z^2}$.
   Find the force as a function of r.

2. **The Potential $V(r) = k/r$**
   The gravitational potential for a point mass M is:

$$V(\vec{r} = k/r) \tag{6.32}$$

where $r = \sqrt{x^2 + y^2 + z^2}$ is the distance from the mass to the point of measurement,[16] and the constant $k = GM$, where G is Newton's constant. Derive the force $\vec{F} = -\nabla V$ in terms of $\hat{r}$ and r.

Problem 3: Del

1. Find ∇V where $V = kx^2$ and k is a constant.

2. Find ∇V where $V = kr^2$ and $r^2 = x^2 + y^2 + z^2$.

3. Find ∇V where $V = kr$ and $r = \sqrt{x^2 + y^2 + z^2}$.

4. Find ∇V where $V = kr^{-1}$ and $r = \sqrt{x^2 + y^2 + z^2}$.

5. Find the scalar product $\nabla \cdot \nabla$ (This is called the Laplacian, ∇^2, after the French mathematical physicist Pierre-Simon Laplace).

6. Find $\nabla^2 r^2$, where $r^2 = x^2 + y^2 + z^2$.

Problem 4: Small Oscillations about a Minimum

Consider a smooth potential in 1-dimension, $V(x)$, that has a minimum $V_0(x_0)$ at $x = x_0$. By expanding $V(x)$ around x_0 in a Taylor series, show that an object trapped in this potential well will execute simple harmonic motion about x_0, and find the frequency.

Problem 5: Behavior of the Simple Pendulum

Prove that a weight on the end of a light string undergoes simple harmonic motion if initially displaced from the vertical.

Problem 6: Numerical Values for a Realistic Pendulum

1. Find the period of a 0.1-kg bob swinging at small angles from the vertical on a rigid pendulum 0.3m long.

2. Find the velocity and acceleration of the bob.

3. Assume the bob is struck horizontally when at rest such that its vertical rise is 2 cm from the rest position. Plot $\theta(t)$, $\dot{\theta}$, and $\ddot{\theta}$ on a single plot versus time.

---

[16] The forces of gravity and electrostatics fall with distance from the source as $1/r^2$, i.e., inverse to the area of a sphere of increasing radius r. The behavior arises from there being no length scale for these forces; the photon and graviton are massless. In contrast, the Strong Interaction (the nuclear force) and the Weak Interaction (responsible for nuclear β-decay, for example), have massive particles as force carriers and consequently have very short ranges.

Problem 7: Another Simple Harmonic Motion Problem: A Mass on a Spring

1. Find the period of a 1-kg mass hanging from a spring with a 10 N/m spring constant under small perturbations.

2. Find the velocity and acceleration of the mass.

3. Assume the mass is held 10cm up from the equilibrium point and released. Plot $x(t)$, $v(t)$, and $a(t)$ on a single plot versus time.

CHAPTER 7

Angular Momentum

7.1   Introduction

The three invariance principles we have invoked in Part I are that the Laws of Physics are invariant under translations in time, in space, and under rotations. Succinctly, the principle is that the equations of physical law are the same in all inertial frames.

Each invariance has an associated conserved quantity. Energy E is the conserved quantity corresponding to invariance under translations in time. The three components p_x, p_y, p_z of momentum $\vec{p}$, are the conserved quantities corresponding to invariance under translations in the three directions in space.

There is another deep-reaching invariance principle, invariance of the laws of physics under rotations, i.e., just as there are no preferred locations or times, there are no preferred directions in space.

In our 3-space there are three axes about which one can rotate, and consequently there are three independent conserved quantities. In Cartesian coordinates these are the angular momenta L_x, L_y and L_z, the components of the vector $\vec{L} = (L_x, L_y, L_z)$ in Cartesian coordinates. No violation of the principle of invariance under rotations has yet been discovered.

7.2   Angular Momentum $\vec{L}$ of a Particle Moving Relative to an Origin

Consider a particle of mass m moving at momentum $\vec{p}$ in a Cartesian coordinate system, at a distance $\vec{r}$ from the origin. The angular momentum of the particle with respect to an axis perpendicular to the plane of motion is defined by the cross product (see Section A.1.1 of Appendix A) of $\vec{r}$ and $\vec{p}$ as shown in Eq. 7.1.

$$\vec{L} = \vec{r} \times \vec{p}. \tag{7.1}$$

With no loss of generality, let us chose motion to be in the xy plane, and the axis about which we calculate the angular momentum consequently to be in the $\hat{z}$

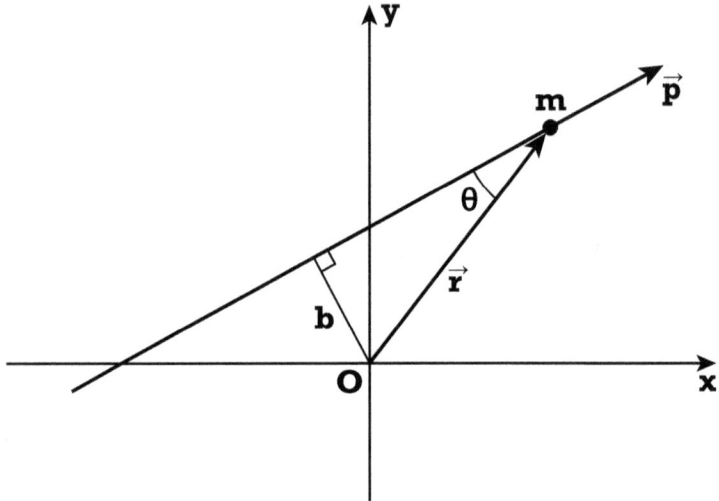

Figure 7.1. The angular momentum of a particle moving in a straight line is proportional to the distance of closest approach to the origin of the frame, defined as the impact parameter b. A particle moving on a trajectory that intersects the origin thus has $L = 0$. Note that in this figure the angular momentum L is in the negative z direction, i.e., into the page.

direction. Figure 7.1 shows the geometry in the xy plane for calculating the angular momentum of a particle at position $\vec{r}$ with momentum $\vec{p}$, measured with respect to the origin. Calculating L_x, L_y, and L_z from equation 7.1:

$$L_x = yp_z - zp_y \quad L_y = zp_x - xp_z \quad L_z = xp_y - yp_x \tag{7.2}$$

and from Fig. 7.1,

$$L_x = 0 \quad L_y = 0 \quad L_z = xp_y - yp_x. \tag{7.3}$$

In index notation:[1]

$$L_i = x_j p_k - x_k p_j \quad \text{and cyclic in } i, j, k. \tag{7.4}$$

7.2.1   Properties of $\vec{L}$

The properties of $\vec{L}$ include:

1. The magnitude is given by $\vec{L} = |\vec{r}||\vec{p}| \sin \theta$, where θ is the angle between $\vec{r}$ and $\vec{p}$.

2. The magnitude of $\vec{L}$ is origin-dependent, and is proportional to the distance of closest approach to the origin, $b = r \sin \theta$, the "impact parameter," i.e., $|\vec{L}| = b|\vec{p}|$ (see Fig. 7.1).

---

[1] I strongly recommend learning index notation; it makes vector algebra so much easier. See Appendix A, Section A.1.1.

3. The magnitude of $\vec{L}$ is proportional to the momentum.

4. $\vec{L}$ is perpendicular to $\vec{r}$ and $\vec{p}$, with the sign determined by the Right Hand Rule (see Section A.1.1 of Appendix A).

5. The dimensionality of angular momentum is $rp = rmv = mr^2/t$, i.e., kg-m$^2s^{-1}$.

7.3   Why Does Angular Momentum Seem Difficult to Grasp?

I consider the physics in this chapter and the next to be the hardest of the course. It is not due to new mathematics—there is nothing new here. But it seems to take the deepest involvement with students in order to get some measure of comfort and confidence.

I have come to the (unwanted) conclusion that the disconnect is intrinsic to our own expectations. The behavior of the objects is different from what we think it should be—in other words, our physical intuition, built from personal experience, is wrong.[2] Having the wrong expectations of the motion is a high barrier to learning the truth.

So you say, "How is it wrong?" The answer is that we naturally expect when we push on an object it goes in the direction we push. As we will learn in this chapter, that is not necessarily so if the object is spinning and we push at an arbitrary point.

For example, a bicycle wheel, detached from the bike, if held by one end of the axle while it spins rapidly in the vertical plane, will not tip over as it would if not spinning, but instead will remain (largely) vertical and will slowly precess around the point of suspension. If stopped, it flops over as expected from the gravitational force at the center-of-mass. What force holds it up when spinning but not when not spinning?

Another non-intuitive behavior is the instability of a spinning object about one of the three possible axes of rotation.[3] Take a rectangular wooden block (or put a rubber band around a book), and flip it into the air spinning around each of the 3 axes of symmetry.[4] For two of the axes, the block will spin around the axis as launched. Around the third, however, the motion is unstable [16]. My own guess is that the disconnect between our intuition and reality stem from the following differences between angular and linear momentum, which is the dominant source of our intuition:

---

[2] Inadequate is gentler, but wrong is more precise. Spinning objects behave in ways that we don't expect and that doesn't feel right.

[3] Remember our preferred formulation of Newton's Second Law, $d\vec{p}/dt = \vec{F}$, and our injunction against working in non-inertial frames.

[4] Flipping a tennis racket about each of the 3 axes is also a traditional demonstration. Try it, it's sweet.

1. $\vec{L}$ is a construct that depends on *two* vectors, $\vec{r}$ and $\vec{p}$, and points along neither;

2. $\vec{L}$ does not lie in the plane defined by the two vectors;

3. $\vec{L}$ depends on the distance to the origin;

4. The change in $\vec{L}$ under an applied force is *perpendicular* to the force.

All of these run contrary to our intuition built up with the vectors of force and momentum, which "act normal."

7.4 $d\vec{L}/dt$ and Torque $\vec{\tau}$

The behavior of the angular momentum of an extended object under the application of a force can be non-intuitive due to the cross-product $\vec{r} \times \vec{p}$ in the definition that makes the change perpendicular to the applied force.[5] Writing $d\vec{L}/dt$ out explicitly:

$$\begin{aligned}
d\vec{L}/dt &= \frac{d}{dt}(r \times \vec{p}) \\
&= (d\vec{r}/dt \times \vec{p}) + (\vec{r} \times d\vec{p}/dt) \\
&= (\vec{v} \times m\vec{v}) + (\vec{r} \times m(d\vec{v}/dt)) \\
&= 0 + (\vec{r} \times m\vec{a})
\end{aligned} \tag{7.5}$$

$$d\vec{L}/dt = \vec{r} \times \vec{F}.$$

This is the definition of torque, denoted by $\vec{\tau}$ (tau):

$$\vec{\tau} = \vec{r} \times \vec{F}. \tag{7.6}$$

The rate of change of the angular momentum $\vec{L}$ is given by:

$$d\vec{L}/dt = \vec{\tau}. \tag{7.7}$$

Note the parallelism in form with $d\vec{p}/dt = \vec{F}$. However, also note the difference; the change of L is always perpendicular to the force.

The left-hand panel of Fig. 7.2 shows a cylindrical shaft with applied forces in the same rotational direction so that the torques add and the forces cancel: $\vec{\tau} = 2R|\vec{F}_{tot}|\hat{z}$, and $\vec{F}_{tot} = 0$. The cylinder will rotate but not translate.

In the right-hand panel, the forces are applied in opposite rotational directions so that the torques cancel and the forces add: $\vec{\tau} = 0$, and $\vec{F}_{tot} = 2\vec{F}$. The cylinder will translate but not rotate.

---

[5] Riders of motorcycles and heavy bicycles will recognize this from the direction of force best applied to turn. Applying a torque by leaning is much more effective than turning the handlebars.

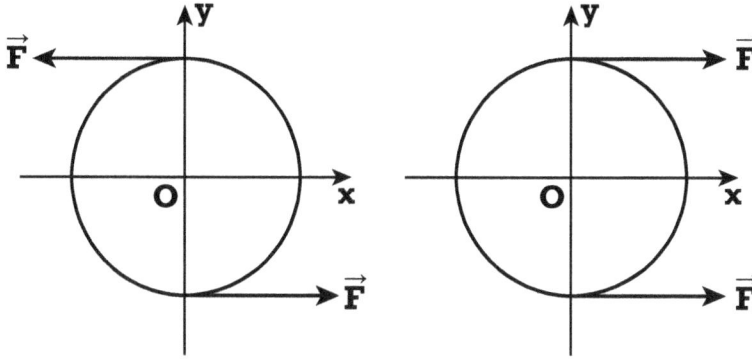

Figure 7.2. A cylindrical shaft with applied forces. Left: The forces are applied in the same rotational direction so that the torques add and the forces cancel: $\vec{\tau} = 2R|\vec{F}_{tot}|\hat{z}$, and $\vec{F}_{tot} = 0$. Right: The forces are applied in opposite rotational directions so that the torques cancel and the forces add: $\vec{\tau} = 0$, and $\vec{F}_{tot} = 2\vec{F}$.

7.5   Velocity and Acceleration in Polar Coordinates

Angular momentum is defined relative to an axis, and consequently we will work in polar coordinates with the axis at the origin.

7.5.1   Rate and Direction of Change of the Unit Vectors

The direction of the unit vectors in polar coordinates $\hat{r}$ and $\hat{\theta}$ depends on the position, unlike the fixed directions of $\hat{x}$ and $\hat{y}$.

Figure 7.3 and Eq. 7.8 show the behavior of the orthogonal unit vectors $\hat{r}$ and $\hat{\theta}$ under a small rotation $\Delta\theta$.

$$\Delta\hat{r} = |\hat{r}|\,\Delta\theta\,\hat{\theta}$$
$$= \Delta\theta\,\hat{\theta}$$
$$\Delta\hat{\theta} = -|\hat{\theta}|\,\Delta\theta\,\hat{r} \tag{7.8}$$
$$= -\Delta\theta\,\hat{r}.$$

Dividing through by Δt and taking the limit $\Delta t \to 0$, we find the radial velocity is given by:

$$\Delta\hat{r} = \Delta\theta\,\hat{\theta}$$
$$\frac{\Delta\hat{r}}{\Delta t} = \frac{\Delta\theta}{\Delta t}\,\hat{\theta} \tag{7.9}$$
$$\frac{d\hat{r}}{dt} = \dot{\theta}\,\hat{\theta}.$$

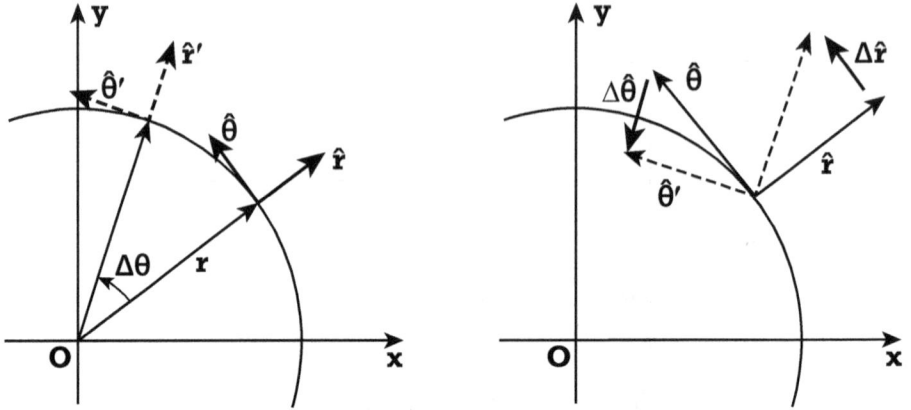

Figure 7.3. Rotation of the axes in polar coordinates. Left: The vectors $\hat{r}$ and $\hat{\theta}$ before an infinitesimal rotation $\Delta\theta$, and after, $\hat{r}'$ and $\hat{\theta}'$. Right: the infinitesimal changes $\Delta\hat{r} = \Delta\theta\hat{\theta}$ and $\Delta\hat{\theta} = -\Delta\theta\hat{r}$. The changes are the numerators of the respective time derivatives: please take good note of the directions and the signs.

The azimuthal velocity is similarly derived:

$$\Delta\hat{\theta} = -\Delta\theta\,\hat{r}$$

$$\frac{\Delta\hat{\theta}}{\Delta t} = -\frac{\Delta\theta}{\Delta t}\,\hat{r} \qquad (7.10)$$

$$\frac{d\hat{\theta}}{dt} = -\dot{\theta}\,\hat{r}.$$

Note that $\frac{d\hat{r}}{dt}$ points in the positive $\hat{\theta}$ direction and $\frac{d\hat{\theta}}{dt}$ points in the *negative* $\hat{r}$ direction.

7.5.2   Velocity in Polar Coordinates

We start with the position vector in polar coordinates:

$$\vec{r} = r\,\hat{r}. \qquad (7.11)$$

Now differentiate with respect to time to get the velocity:

$$\vec{v} = \frac{d\vec{r}}{dt}$$

$$= \frac{dr}{dt}\,\hat{r} + r\,\frac{d\hat{r}}{dt} \qquad (7.12)$$

$$\vec{v} = \dot{r}\,\hat{r} + r\dot{\theta}\,\hat{\theta}. \qquad (7.13)$$

Note that the velocity in the $\hat{\theta}$ direction is $r\dot{\theta}$, and thus increases linearly with radius at constant $\dot{\theta}$.

7.5.3   Rotation at Constant Angular Velocity: ω (Omega)

For constant angular velocity, we define $\omega \equiv \dot{\theta}$, measured in radians/sec, i.e., with dimensions of inverse time. As seen directly from Eq. 7.10, the unit vectors change perpendicular to their direction linearly with ω and with opposite signs:

$$\frac{d\hat{r}}{dt} = \omega\,\hat{\theta}$$
$$\frac{d\hat{\theta}}{dt} = -\omega\,\hat{r}. \tag{7.14}$$

Note that since they are unit vectors, the magnitudes do not change.

7.5.4   Acceleration in Polar Coordinates

To find the acceleration, we differentiate once more with respect to time:

$$\vec{a} = \frac{d\vec{v}}{dt}$$
$$= \frac{d(\dot{r}\,\hat{r} + r\dot{\theta}\,\hat{\theta})}{dt} \tag{7.15}$$
$$= \ddot{r}\,\hat{r} + \dot{r}\,\frac{d\hat{r}}{dt} + \dot{r}\dot{\theta}\,\hat{\theta} + r\ddot{\theta}\,\hat{\theta} + r\dot{\theta}\,\frac{d\hat{\theta}}{dt}$$
$$= \ddot{r}\,\hat{r} + \dot{r}\dot{\theta}\,\hat{\theta} + \dot{r}\dot{\theta}\,\hat{\theta} + r\ddot{\theta}\,\hat{\theta} - r\dot{\theta}^2\,\hat{r}.$$

Collecting terms in $\hat{r}$ and $\hat{\theta}$:

$$\vec{a} = (\ddot{r} - r\dot{\theta}^2)\,\hat{r} + (r\ddot{\theta} + 2\dot{r}\dot{\theta})\,\hat{\theta}. \tag{7.16}$$

7.6   Centripetal Force and the Coriolis Effect

The acceleration of Eq. 7.16 has two terms that mix the two coordinates r and θ. These terms are $(-r\dot{\theta}^2)$ in the $\hat{r}$ direction, and $2\dot{r}\dot{\theta}$) in the $\hat{\theta}$ direction. We discuss these in turn below.

7.6.1   Centripetal Force

Consider the component of the acceleration in the radial direction in Eq. 7.16. The first term, $\ddot{r}$, is the familiar acceleration due to the second derivative of $\vec{r}$. The second term, $-r\dot{\theta}^2$, however, may be unfamiliar: it points at the origin and is proportional to $\dot{\theta}^2$. This acceleration is due to an applied Centripetal Force that keeps the object moving on a locally circular path (see Section 7.7). The colloquial expression "Centrifugal Force" refers to the centripetal acceleration as seen in the non-inertial frame. As an example, consider holding a cup of coffee as a passenger in a car as the car takes a hard left turn. The coffee wants to go "straight" (Newton's First Law); the cup exerts a transverse force to change the coffee's momentum to the left.

7.6.2   The Coriolis Effect

The azimuthal component of the acceleration in Eq. 7.16 is

$$a_\theta = (r\ddot{\theta} + 2\dot{r}\dot{\theta}).$$ (7.17)

The first term is the acceleration due to the second derivative of the angle with time. The second term, however, has two factors of time in the denominator,[6] not from a second derivative, but instead from the product of a first derivative in r and a first derivative in θ. This term is known as the Coriolis "Force," or better, as the Coriolis Effect, as it is not a force. The effect is due to the change in radius and conservation of angular momentum, as illustrated in Figure 7.4

In the (non-inertial!) frame of the Earth, the trajectory is seen to behave under a sideways force. However, to an observer in a fixed inertial frame in which the Earth is rotating, the trajectory stays in a vertical plane.

7.7   Circular Motion

In circular motion, $\dot{r} = 0$, and so Eq. 7.13 reduces to:

$$\vec{v} = r\dot{\theta}\,\hat{\theta}$$ (7.18)

At each point on the circle the velocity is tangential and, in our convention, counter-clockwise as shown in Figure 7.5.

7.7.1   Uniform Circular Motion

For uniform circular motion, $\dot{\theta}$ is a constant, defined as $\omega \equiv \dot{\theta}$. The acceleration points at the center of the circle, and is given by

$$\vec{a} = -r\dot{\theta}^2\hat{r} = -r\omega^2\hat{r}.$$ (7.19)

---

[6] Dimensional analysis yet again.

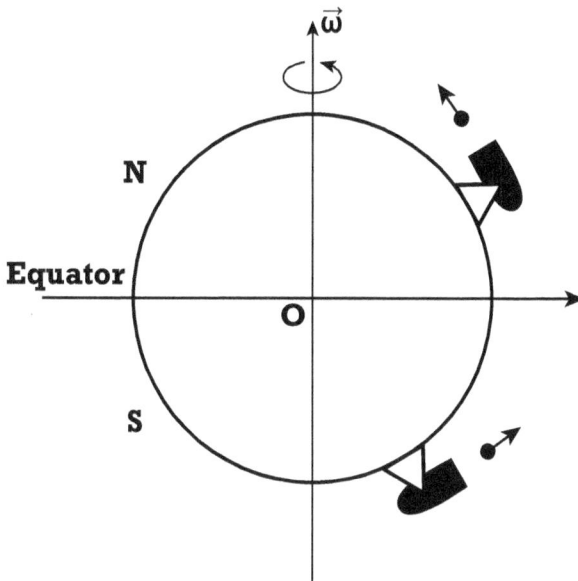

Figure 7.4. A visualization of the Coriolis effect from the myth of the naval engagement at the Falklands Islands between England and Germany in World War I. A cannon fires a shell towards the North in each of the Northern and Southern hemispheres. In the Northern hemisphere, the shell moves closer to the axis of the Earth's rotation, so to conserve angular momentum it speeds up relative to the Earth, and so will land East of the target (the sun comes up in the East, so the Earth must rotate West to East). The shell is rotating faster than the Earth. In the Southern hemisphere, the shell moves farther from the Earth's rotation axis, and so slows down relative to the Earth, which rotates underneath it so that the shell will land West of the target. However, none of this was relevant to the battle [33].

The angular frequency ω is related to the frequency of revolution ν by[7]

$$\omega = 2\pi\nu. \tag{7.20}$$

7.7.2   The Golden Rule of Circular Motion

Seemingly complex problems such as the minimum height a roller-coaster car must start at to do a loop-the-loop without falling off the track (see Problem 7.6) yield to a simple rule:

If a particle is moving on a locally circular path of radius R, there is a centripetal force $F = mv^2/R$ pointing inwards. If there is a force $F = mv^2/R$ pointing inwards then the particle is moving on a locally circular path.

---

[7] I remember this by the mnemonic: "ω is larger than ν" as it has two squiggles and ν only one.

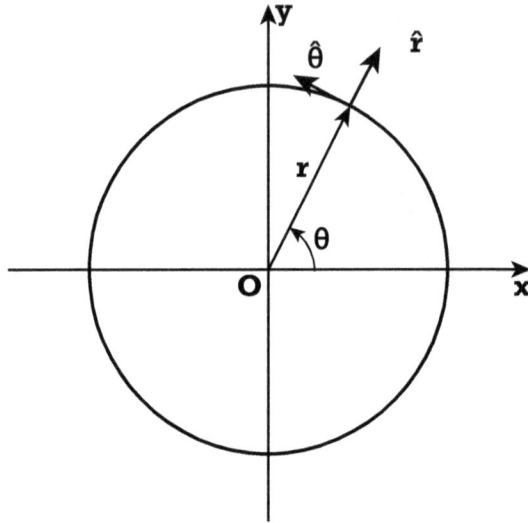

Figure 7.5. Circular motion. The unit vector $\hat{r}$ points radially, and the unit vector $\hat{\theta}$ points tangentially, at every point on the circle.

7.7.3   The Fictitious Centrifugal Force

Much (too much) is made in physics textbooks of the common use of the phrase "centrifugal force." Suffice it to say that for locally circular motion analyzed in an inertial frame[8] there is an acceleration caused by an applied force pointing to the center of the circular motion, aptly named the "centripetal" force. For example, the centripetal force is the force of the string on an object whirled in a circle, the force of the spoke on a rim of a bicycle wheel pulling that part of the tire into a circular trajectory when the wheel turns, the force of the seat on you when the car you are riding in turns. There is nothing "fictitious" about centripetal forces.

7.7.4   Angular Momentum in Polar Coordinates

$$
\begin{aligned}
\vec{L} &= \vec{r} \times \vec{p} \\
&= \vec{r} \times m\vec{v} \\
&= r\,\hat{r} \times m(\dot{r}\,\hat{r} + r\dot{\theta}\,\hat{\theta}) \\
&= r\,\hat{r} \times m(\dot{r}\,\hat{r}) + mr\,\hat{r} \times (r\dot{\theta}\,\hat{\theta}) \\
&= mr\dot{r}\,(\hat{r} \times \hat{r}) + mr^2\dot{\theta}(\hat{r} \times \hat{\theta}) \\
&= \qquad 0 \quad + mr^2\dot{\theta}(\hat{r} \times \hat{\theta}) \\
\vec{L} &= (mr^2\dot{\theta})\,\hat{z}.
\end{aligned}
\tag{7.21}
$$

---

[8] Remember the injunction: "Just say NO" to working in non-inertial frames.

For uniform circular motion about the z axis

$$|L| \equiv L = mr^2\omega \tag{7.22}$$

and L points along the positive z axis.

Because L is a constant of the motion if there is no torque, we will use this relationship to eliminate ω in the acceleration in circular motion (Eq. 7.19): Solving for ω:

$$\omega = L/mr^2 \tag{7.23}$$

and the force term in the equation of motion becomes independent of ω:

$$ma = -mr\,\dot\theta^2$$
$$= -mr\,\omega^2 \tag{7.24}$$
$$= L^2/(mr^3).$$

We will take advantage of L being a constant of the motion to simplify the Kepler Problem in Chapter 9.

7.8   Problem Set 7: Angular Momentum; Torque, Velocity and Acceleration in Polar Coordinates, Circular Motion

Study Groups: You *must* work collaboratively with a functioning study group. However, the work you hand in **has to be your own**.

Formulae for Velocity and Acceleration in Polar Coordinates:
There should be little memorization in this course. However, the change in direction of the unit vectors and the representations of velocity and acceleration in polar coordinates are tedious to reconstruct. For the duration of the course it's worth remembering:

$$\frac{d\hat r}{dt} = \dot\theta\,\hat\theta$$

$$\frac{d\hat\theta}{dt} = -\dot\theta\,\hat r \tag{7.25}$$

$$\vec v = \dot r\,\hat r + r\dot\theta\,\hat\theta$$

$$\vec a = (\ddot r - r\dot\theta^2)\,\hat r + (r\ddot\theta + 2\dot r\dot\theta)\,\hat\theta.$$

Problem 1: Cross Products

Explicitly calculate the following cross-products. For lines 6 and 7 you may want to use the Levi-Civita tensor (Appendix A, Section A.1.8) to calculate one component.

1. $\hat{x} \times \hat{y}$; $\hat{y} \times \hat{z}$; $\hat{z} \times \hat{x}$
2. $\hat{x} \times \hat{z}$; $\hat{y} \times \hat{x}$; $\hat{z} \times \hat{y}$
3. $(y\hat{x} - x\hat{y}) \times (y\hat{x} - x\hat{y})$
4. $(y\hat{x} - x\hat{y}) \times (z\hat{x} - x\hat{z})$
5. $\vec{A} \times \vec{B}$
6. $\vec{A} \times (\vec{B} \times \vec{C})$
7. $(\vec{A} \times \vec{B}) \times (\vec{B} \times \vec{C})$

Problem 2: Torque

Starting with the equation for the time dependence of $\vec{L}$:

$$d\vec{L} \,/dt \tag{7.26}$$

derive the definition of torque:

$$\tau \equiv d\vec{L} \,/dt = \vec{r} \times \vec{F}. \tag{7.27}$$

Problem 3: Velocity and Acceleration in Polar Coordinates

Starting with the position vector in polar coordinates

$$\vec{r} = r\hat{r} \tag{7.28}$$

derive the equations in polar coordinates for:

1. velocity;
2. acceleration.

(Best done without looking at the book or elsewhere).

Problem 4: Angular Momentum in Polar Coordinates

Derive the expression for the angular momentum in polar coordinates:

$$\vec{L} = (mr^2\dot{\theta})\hat{z}. \tag{7.29}$$

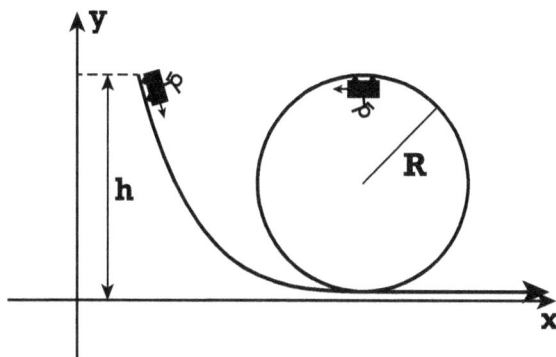

Figure 7.6. Problem 5. A roller-coaster with a circular section in which the cars travel in a full circle so that at the top the passengers are upside-down. The car starts from rest at height h; the idea is that it doesn't fall off the track.

Problem 5: The Loop-the-Loop Roller Coaster

Consider the roller coaster Loop-the-loop, in which the car has enough velocity to go around a circle in the vertical plane without falling off the tracks, as shown in Figure 7.6. Ignoring friction and air resistance, determine the minimum height for the starting point for the car so it doesn't fall (a prize for the shortest derivation in your study group and in the class?).

Problem 6: Centripetal Force in the Loop-the-Loop Roller Coaster

For the roller coaster in Problem 5 and Figure 7.6, find the vertical force on the car as a function of position on the track. As in Problem 5, assume the car has just enough velocity to go around a circle in the vertical plane without falling off the tracks. Please draw and label your coordinate system including showing an example position of the car (i.e., on the track somewhere other than at the top or bottom).

Problem 7: Circular Motion and SHM (harder)

A bead is put on a circular wire hoop so that the bead is constrained to move on the wire as shown in Fig. 7.7. The hoop is rotated about its vertical axis.

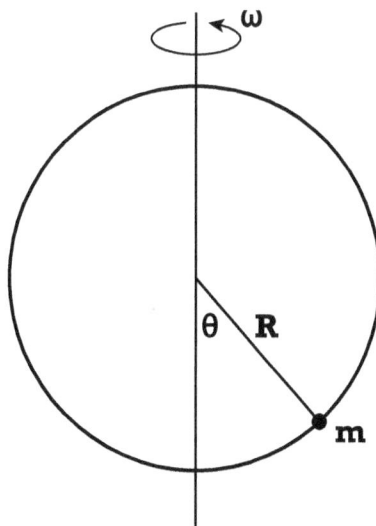

Figure 7.7. Problem 7. A bead sliding on a circular wire hoop rotating about the vertical axis.

1. Find the stable position of the bead as a function of the angular frequency of rotation.

2. Find the frequency of oscillations about the stable point as a function of the angular frequency of rotation.

Problem 8: Challenge Problem—Tower of Song

(I suggest discussing this with your study group before beginning to calculate. Knowing what frame you are in and drawing a Free-Body Diagram will help a lot.)

Leonard Cohen is visiting the music scene on the South Side of Chicago and takes the Metra Electric train to the Loop downtown. While still at the 57th Street station, Leonard sets his bowling ball on the floor of the car in the aisle next to his seat. The train starts to move with constant acceleration a. Find the acceleration of the ball versus time, assuming it rolls without slipping.

CHAPTER 8

Rigid Bodies, the Moment of Inertia Tensor, Collisions and Chasles' Theorem

8.1   Introduction

A rigid body is an extended object in which the relative distance between any two points is fixed, and so the internal mass distribution does not change under the application of forces. Equivalently, we assume that the object has no internal degrees of freedom.

To characterize the body independent of any rotation, we define a new frame of reference, the body frame, to be a Cartesian coordinate system with origin at the center-of-mass (CM) and axes fixed in the body. The mass distribution in the body frame does not change when the body rotates in the lab frame. Our convention is that the axes in the body frame and the axes in the lab frame line up at $t = 0$.

We first treat collisions of a particle with a rigid body, working in the lab frame to calculate the linear motion of the CM and the rotation of the body about the CM. The linear motion is simple: the motion of the CM of the object is that of a point particle of the same mass that obeys conservation of momentum. The rotation is equally simple once the moment of inertia is calculated; the angular momentum of the object around an axis obeys conservation of angular momentum about the CM.

We then calculate the moment of inertia tensor $\vec{\vec{I}}$ in the general case; $\vec{\vec{I}}$ is a 3×3 symmetric matrix, with 6 independent matrix elements.[1]

We conclude with several common examples of the axially-symmetric case, a ring and a disk, to firmly (but gently) convey the r^2 dependence of the moment on the distance to the axis of rotation for the diagonal elements of $\vec{\vec{I}}$.

---

[1] By symmetry the off-diagonal elements are equal, and so there are 6 and not 9 independent elements: 3 diagonal plus 3 off-diagonal.

8.2   Chasles' Theorem

The seemingly complex motion after a collision involving a particle and a rigid body has a remarkable simplification, known as Chasles' Theorem. We develop the tools to discuss it below.

8.2.1   Conservation of Momentum of the CM and Conservation of Angular Momentum about the CM

The discussion of collisions in Chapter 4 was for point-masses; extending the analysis to rigid bodies introduces additional rotational degrees of freedom. Happily there are additional equations of constraint, as angular momentum is conserved. This leads to an elegant solution to a collision of a point particle with a rigid body: the linear and rotational motions of the system separate into constraints on the momentum of the center-of-mass (CM) in the lab frame, and on angular momentum around the CM, respectively. This is called Chasles' Theorem [15].[2]

Chasles' Theorem states:

1. The CM of the system acts like a point particle obeying Newton's Second Law under the total force on the system:

$$\frac{d\vec{p}}{dt} = \sum_i \vec{F}_i. \qquad (8.1)$$

2. The rotation *about the CM* is described by the Moment of Inertia tensor multiplying an angular velocity vector:

$$\vec{L} = \overset{\leftrightarrow}{I}\vec{\omega} \qquad (8.2)$$

where $\overset{\leftrightarrow}{I}$ is a 3×3 symmetric matrix calculated in a coordinate system that rotates with the object (the body frame), and the angular momentum $\vec{L}$ and angular frequency $\vec{\omega}$ are vectors in the lab CM frame. Figure 8.1 shows an example rigid body and its imbedded body frame coordinates that rotate with the body about the CM.

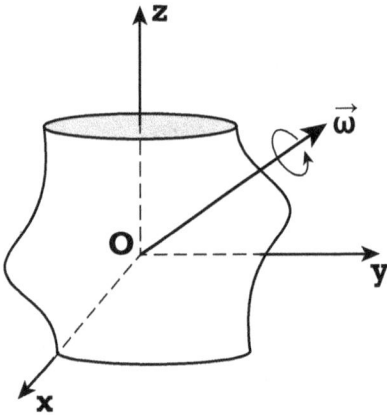

Figure 8.1. An asymmetric rigid body, rotating about an arbitrarily chosen axis through its center-of-mass at angular velocity $\vec{\omega}$. The axes of the body frame rotate about the origin in the lab frame; the origin of the body frame is the center-of-mass (CM). The motion of the CM in the lab frame is that of a point particle with the mass of the body. This separation is known as Chasles' Theorem.

---

[2] The name does not seem to be widely known, perhaps because of a difficult history and derivation.

8.3   Matrix Relationship of $\vec{L}$ and $\vec{\omega}$

As discussed in Section 8.4, an axially-symmetric object spinning about the axis of symmetry has $\vec{L}$ parallel to $\vec{\omega}$, $\vec{L} = I\vec{\omega}$, where I is a scalar. However, in real life, a rigid body will not have perfect axial symmetry.[3]

8.3.1   The Two Frames: The Rotating Body Frame and the Fixed Lab CM Frame

Chasles' Theorem tells us that we can separate the motion of a rigid body into the motion of the center-of-mass (CM), which obeys Newton's Second law under the total impressed force, and rotation about the CM, which obeys conservation of angular momentum.[4] We define a Cartesian coordinate system fixed in the body with the origin at the CM, as shown in Figure 8.1, that rotates with the body relative to the laboratory CM frame. The geometry of the rigid body consequently does not change with time in the body frame, and depends only on the mass distribution (shape and/or density variations) of the body.

8.3.2   The Moment of Inertia Tensor $\vec{\vec{I}}$

Writing out $\vec{L} = \vec{\vec{I}}\,\vec{\omega}$ explicitly:

$$\begin{pmatrix} L_x \\ L_y \\ L_z \end{pmatrix} = \begin{pmatrix} I_{xx} & I_{xy} & I_{xz} \\ I_{yx} & I_{yy} & I_{yz} \\ I_{zx} & I_{zy} & I_{zz} \end{pmatrix} \begin{pmatrix} \omega_x \\ \omega_y \\ \omega_z \end{pmatrix}. \tag{8.3}$$

Thus L_x, for example, has contributions from ω_y and ω_z:

$$L_x = I_{xx}\,\omega_x + I_{xy}\,\omega_y + I_{xz}\,\omega_z \tag{8.4}$$

where L_x is the x-component of the angular momentum and ω_x is the x-component of the angular velocity both in the lab CM frame, and the matrix elements of $\vec{\vec{I}}$ are calculated in the body frame.

---

[3] Even precision-machined bearings and axles have a finite precision. Most things in daily life, such as bicycle wheels with valve stems, car wheels balanced on a local balancing machine, and footballs with seams, have perceptible asymmetries, while many common objects, such as tennis rackets, silverware, and dishes are wildly asymmetric about at least one axis [16].

[4] That is, the angular momentum before the collision, which is all in the incoming particle, is equal to the total angular momentum after the collision.

8.3.2.1   Dimensionality of the Matrix Elements of $\vec{\vec{I}}$

Angular momentum L is proportional to $|\vec{p}|\vec{r}| \propto mvr$, and so has units of kg-m^2/s. Since angles are dimensionless,[5] ω has units of s^{-1}. Moments of inertia consequently have dimensions of mass $\times$ length-squared, i.e., kg-m^2.

Physically the r^2 dependence makes sense. For a given angular frequency ω, the velocity of a mass element goes as the distance r to the axis. An additional power of r comes from the r in the angular momentum mvr.

8.3.3   Angular Momentum and Moments of Inertia in Index Notation

In compact index notation, where i and j each run from 1 to 3, Equation 8.4 becomes:

$$L_i = \sum_{j=1}^{3} I_{ij}\,\omega_j. \tag{8.5}$$

Using the convenient Einstein summation convention (Section A.1.9 of Appendix A), that a repeated index, in this case the index j, is always summed over:[6]

$$L_i = I_{ij}\,\omega_j. \tag{8.6}$$

8.3.4   Kinetic Energy of Rotation, T

We expect the kinetic energy to be quadratic in velocity, and so will have a dependence that goes as r^2 and ω^2. The r^2 factor is incorporated in the moment of inertia. The dimensions of kinetic energy[7] are those of $\frac{1}{2}mv^2$, i.e., kg-m^2-s^{-2}.

The kinetic energy for the general case is given by the sum:

$$T = \frac{1}{2}I_{ij}\omega_i\omega_j \tag{8.7}$$

where ω_i is the angular frequency about the i axis.

The kinetic energy of an axially-symmetric rigid body rotating at angular velocity ω around the axis of symmetry is

$$T = \frac{1}{2}I\omega^2 \tag{8.8}$$

where I is the (scalar) moment of inertia about the axis of rotation (see Section 8.4).

Note the spatial dimensionality: the moment of inertia has two powers of r, which combine with the 2 powers of ω to give a velocity-squared dependence.

---

[5] Angles are determined by ratios of lengths.

[6] You will get used to not writing the summation symbol and limits (lots of strokes!)—Einstein realized they can be dropped with no loss of information.

[7] It's unfortunate that the units of length are meters, and so m here stands for a unit of length, meters, and not mass, which is kg. Apologies—we eventually learn to live with these booby traps.

8.3.5 Calculating the Matrix Elements of the Inertia Tensor

The 6 independent matrix elements of the inertia tensor are sufficient to describe the relationship between $\vec{L}$ and the angular frequency $\vec{\omega}$ for all rigid bodies. To illustrate the calculation of the 3 diagonal and 3 off-diagonal elements in the general case we will consider a specific example, a small[8] mass element of a rigid body rotating about the z-axis. The mass of the element is $\rho d^3 V = \rho(x,y,z)\,dxdydz$, where $\rho(x,y,z)$ is the density at the point x,y,z. We first calculate the diagonal elements.

8.3.5.1 The Diagonal Elements of $\overset{=}{I}$

To make this less abstract, a little volume element of dimensions $dx = 1$ mm, $dy = 1$ mm, $dz = 1$ mm, of a solid steel object has a mass $dm = (0.1\ \text{cm})^3 (8\ \text{gm/cm}^3) = 8 \times 10^{-3}$ gm $= 8$ milligrams.[9] Assume the mass element rotates about the z axis 1 cm from the axis:

$$I = mr^2$$

$$= (8 \times 10^{-3}\ gm)\ (1\ cm)^2$$

$$= (8 \times 10^{-3}\ gm)\ (\frac{1\ kg}{10^3\ gm})\ ((1\ cm)\ (\frac{1\ m}{100\ cm}))^2 \qquad (8.9)$$

$$= 8 \times 10^{-8}\ kg\text{-}m^2$$

where we have used the identities $1 \equiv (\frac{1\ kg}{10^3\ gm})$ and $1 \equiv (\frac{1\ m}{100\ cm})$ to convert grams to kilograms and centimeters to meters.[10]

Rotating about the z-axis contributes to L_z by:

$$\begin{pmatrix} L_x \\ L_y \\ L_z \end{pmatrix} = \begin{pmatrix} I_{xx} & I_{xy} & I_{xz} \\ I_{yx} & I_{yy} & I_{yz} \\ I_{zx} & I_{zy} & d^3mr^2 \end{pmatrix} \begin{pmatrix} 0 \\ 0 \\ \omega_z \end{pmatrix}. \qquad (8.10)$$

The matrix element I_{zz} connects L_z to ω_z, and since in our example ω points only in the z direction,

$$L_z = I_{zz}\,\omega_z$$

$$= (d^3 m)\ r^2\ \omega_z \qquad (8.11)$$

$$= (d^3 m)\ (x^2 + y^2)\ \omega_z.$$

---

[8] The scale of "small" is set by the length over which the mass distribution can be considered changing linearly.

[9] Math majors always object to the use of the symbol for a differential quantity, dm, for a finite small quantity. It doesn't matter.

[10] In unit conversion, to determine which unit should be in the numerator and which in the denominator, note that we want to convert kg in the numerator to g, so when multiplying by the unit conversion, kg should be in the denominator and g in the numerator, i.e., we multiply by $1 \equiv 1000\ g/kg$.

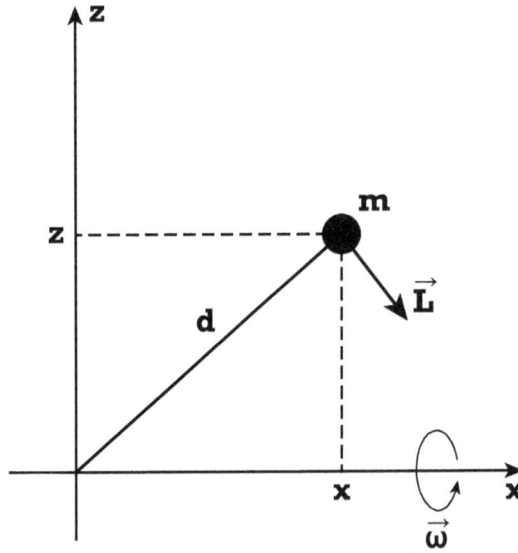

Figure 8.2. A mass m spinning about the x axis. The angular momentum $\vec{L}$ has components in the x and y directions when the rotator is in the xy plane.

To integrate this over the entire body we replace the little mass element d^3m with the local density times volume element, $\rho(x, y, z)\, d^3V$, with units of $(\text{kg/m}^3)(\text{m}^3)=$ kg. Factoring out the common factor of ω_z, we have

$$I_{zz} = \int \int \int \rho(x^2 + y^2)\, d^3V. \tag{8.12}$$

Suppose we had instead rotated about the x axis with the same d^3m. We can invoke the cyclic nature (See Appendix A, Section A.1.8) of a right-handed Cartesian system to replace z with x, x with y, and y with z:

$$
\begin{aligned}
I_{xx} &= \int \int \int \rho\,(y^2 + z^2)\, d^3V \\
I_{yy} &= \int \int \int \rho\,(z^2 + x^2)\, d^3V
\end{aligned}
\tag{8.13}
$$

In each case the quantity under the integral is the product of an infinitesimal mass element d^3m and its distance-squared to the axis, r^2.

8.3.5.2 The Off-Diagonal Elements of $\overset{\leftrightarrow}{I}$

Consider the one-arm Tinkertoy rotator spinning around the x axis as shown in the xz plane in figure 8.2. As shown in the figure, the angular momentum $\vec{L}$ lies in the xz plane at $t = 0$, and so has both L_x and L_z non-zero components.

The L_z component arises from a velocity with magnitude $z\omega$ around the x axis at a distance x from the z axis. The product of velocity times lever arm is $-mxz\omega$.

By our convention for $\vec{\vec{I}}$:

$$L_z = I_{zx}\,\omega_x \tag{8.14}$$

$$I_{zx} = -mxz. \tag{8.15}$$

The minus sign arises from the clockwise (positive) velocity around the x axis leading to a counter-clockwise motion (negative) around the z-axis (see the direction of $\vec{L}$ in Figure 8.2).

Generalizing to the ij off-diagonal element of the tensor,

$$I_{ij} = -m\,x_i x_j. \tag{8.16}$$

8.3.5.3 The Moment of Inertia $\vec{\vec{I}}$ Is a Symmetric Matrix with Six Independent Elements

Ab initio one would think that the 3 matrix $\vec{\vec{I}}$ would have nine independent elements. However, Equation 8.16 is symmetric with respect to i and j, with $I_{ij} = I_{ji}$. The Moment of Inertia Tensor $\vec{\vec{I}}$ is consequently a symmetric matrix. In addition to the three diagonal elements M_{ii} there are three independent off-diagonal elements M_{ij}, for a total of six parameters to fully describe the angular momentum and kinetic energy of a rotating rigid body.

8.3.6 Summary of the Matrix Elements of $\vec{\vec{I}}$ in Tensor Notation

We can summarize the components of the moment of inertia matrix succinctly as:

$$I_{ij} = \int \int \int \rho\,(\vec{r})\,d^3r\,[\delta_{ij}x_k x_k - x_i x_j] \tag{8.17}$$

where i, j and k run from 1 to 3, and by convention typically represent x, y, and z respectively, and δ_{ij} is the Kronecker delta (see Appendix A, Section A.1.8).

Checking Equation 8.17 for both a diagonal and off-diagonal element:

$$I_{11} = \int \int \int \rho(\vec{r})\,d^3r\,[(x_1^2 + x_2^2 + x_3^2) - x_1^2]$$

$$= \int \int \int \rho(\vec{r})\,d^3r\,[(x_2^2 + x_3^3)] \tag{8.18}$$

$$I_{12} = \int \int \int \rho(\vec{r})\,d^3r\,[-x_1 x_2].$$

The left-hand panel of Figure 8.3 shows a working diagram for the Tinkertoy rotator. The centripetal force on each mass exerted by the supporting rod corresponds to a torque $\vec{r} \times \vec{F}$ perpendicular to the rod and to the radial vector from the axis to the mass. This torque is countered by the constraining force on the vertical shaft of the rotator. The right-hand panel shows a rotator symmetric around the vertical axis; in this case there is no net torque from the two masses, and $\vec{L}$ is "parallel to" $\vec{\omega}$.

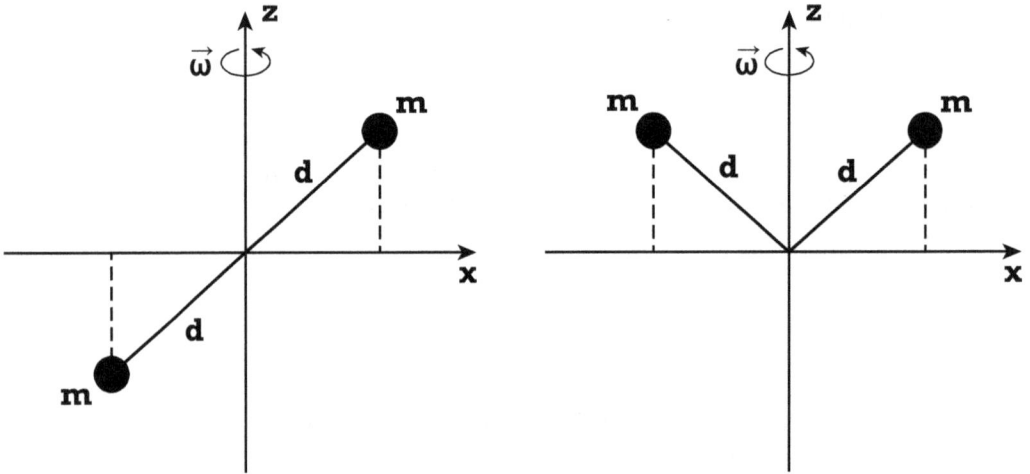

Figure 8.3. Left: The tilted Tinkertoy rotator. Two masses rotate about the vertical axis. Each is held in place by a supporting rod out of the horizontal plane. The centripetal force on each mass exerted by the supporting rod generates a torque $\vec{\tau} = \vec{r} \times \vec{F}$ perpendicular to the rod and to the radial vector from the axis to the mass. To determine the sign of τ, in the absence of the torque the masses would move further from the axis to lie in the horizontal plane. The torque is countered by the constraining force on the vertical shaft of the rotator. Right: a rotator symmetric about the vertical axis.

8.4 The Moment of Inertia I for a Uniform Axially-Symmetric Body Rotating about the Axis of Symmetry

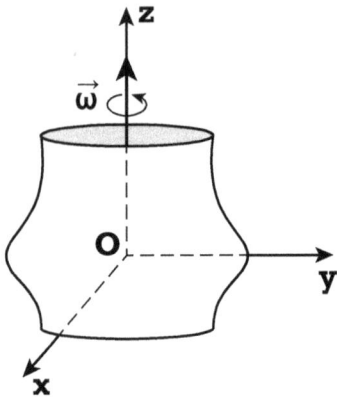

Figure 8.4. An axially-symmetric rigid body of radius R rotating at angular velocity ω about the z axis.

Consider the uniform axially-symmetric body of mass M and radius R rotating about the axis of symmetry with angular velocity ω shown in Figure 8.4. By symmetry, the angular momentum points in the same direction as ω along the z axis:[11]

$$\vec{L} = I\vec{\omega} \qquad (8.19)$$

where I is a scalar.

Because the body is axially-symmetric and rotating about the z-axis, only the radial coordinate will enter into the calculation of I, and we have to integrate only over that one variable.

---

[11] Our convention is that "right-handed"—i.e., the direction of fingers on the right hand when the thumb is along the axis—has positive ω.

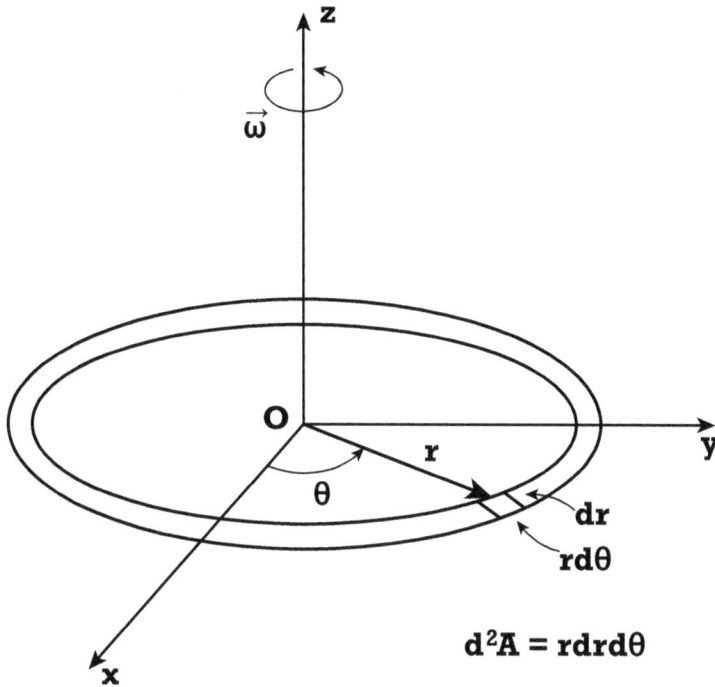

Figure 8.5. A uniform ring of mass M and radius R rotating at angular velocity ω about the z axis.

The moment of inertia for an axially-symmetric body is found by (see Appendix A, Section A.1.5):

$$I = \int_0^R \rho(r) \, r^2 dr \qquad (8.20)$$

where $\rho(r)$ is the mass between r and $r + dr$ as a function of r.

8.4.1   Calculating the Moment of Inertia for Some Common Axially-Symmetric Bodies

8.4.1.1   Calculating I for a Ring

Figure 8.5 shows the body frame for a ring of radius r and radial thickness dr. We assume the ring is homogenous and $dr << r$, i.e., it's a thin ring. Since the moment I depends only on the radius from the axis of rotation and the ring is thin in the z direction, we can treat this as a problem in two dimensions.

First, let's do it by brute (brute intellectual?) force. For a ring of radius R, all the mass is at $r = R$, and so there is no need to calculate the integral for the moment:

$$I = \int_0^R \rho(r) \, r^2 dr \qquad (8.21)$$

as we know that for a single mass M at radius R the moment is $I = MR^2$.

As all the mass of the ring is also at radius R, the moment about the axis of symmetry for a ring of mass M at radius R is:

$$I = MR^2. \tag{8.22}$$

Define a 1-dimensional mass density λ per unit length around the ring.[12]

$$\lambda = \frac{M}{2\pi R}. \tag{8.23}$$

The differential mass element is:

$$dm = \lambda \, Rd\theta. \tag{8.24}$$

Integrating the mass around the ring:

$$\begin{aligned}
I &= \int_0^{2\pi} (\lambda \, Rd\theta) \, R^2 \\
&= \lambda \, R^3 \int_0^{2\pi} d\theta \\
&= \frac{M}{2\pi R} (R^3) (2\pi) \\
I &= MR^2.
\end{aligned} \tag{8.25}$$

8.4.1.2 Calculating I for a Uniform Circular Disk about the Axis

We can use Eq. 8.25 for a ring to calculate the moment for a thin uniform circular disk around its axis. Figure 8.6 shows a uniform disk of radius R and mass per unit area $\sigma = M/(\pi R^2)$, rotating about the axis of symmetry. We can solve for I for the disk by integrating rings of area $2\pi r dr$ and mass density σ, weighted by r^2, from 0 to R:

$$\begin{aligned}
I &= \int_0^R \sigma \, (2\pi \, rdr) \, r^2 \\
&= 2\pi\sigma \int_0^R r^3 dr \\
&= 2\pi \frac{M}{\pi R^2} \frac{R^4}{4} \\
&= \frac{1}{2} \frac{M}{R^2} R^4 \\
I &= \frac{1}{2} MR^2.
\end{aligned} \tag{8.26}$$

---

[12] A common convention is that ρ (rho) is a density in 3-dimensions, σ (sigma) is a density in 2-dimensions, and λ (lambda) is a density in 1-dimension.

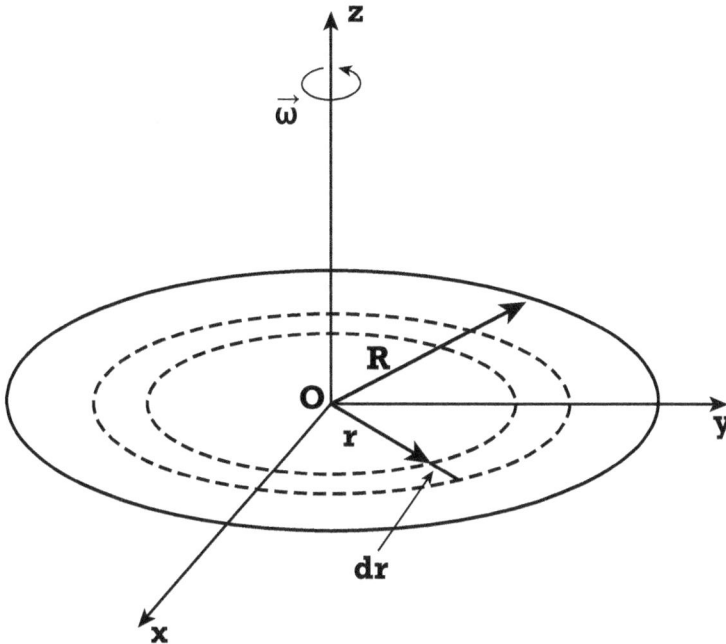

Figure 8.6. A uniform disk of mass M and radius R rotating at angular velocity ω about the z axis. The moment of inertia is calculated in Eq. 8.26 by integrating the moment for rings of smaller radius r, shown as dashed lines and calculated in Eq. 8.25.

The moment of a disk being half that of a ring corresponds to the average value of $\sin^2\theta$. This factor comes from the distance-squared from a mass element to the axis of rotation.

8.4.1.3 *I for a Uniform Axially-Symmetric Cone about the Axis*

Consider a uniform axially-symmetric cone centered on the z axis. In the same spirit that we built a disk out of concentric rings, using a solution in 1-D to solve a problem in 2-D, we can consider the cone to be an infinite stack of infinitely thin disks of decreasing radius. We leave solving for I for the cone to the Problem Set.

8.5   Problem Set 8: Chasles' Theorem, Rigid Bodies, the Moment of Inertia Tensor

Problem 1: Chasles' Theorem

A snowball sliding on an icy pond hits near the end of a board and sticks to the board's side, as shown in Figure 8.7. The motion of the system after the

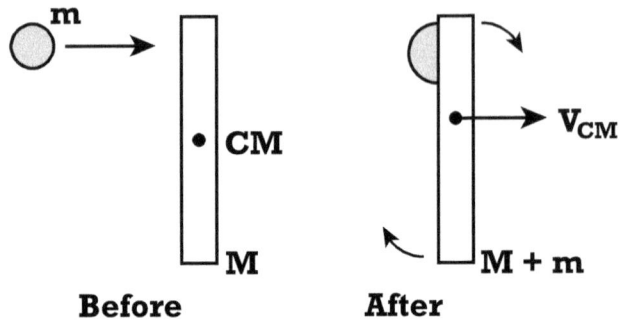

Figure 8.7. Problem 1. The inelastic collision of a snowball sliding across an icy pond and a board lying on the ice. The snowball hits at one end of the board and sticks to it. The motion of the CM of the combined system in the lab frame is determined by conservation of momentum, and the rotation of the system about the CM in the CM frame by conservation of angular momentum. This separation is known as Chasles' Theorem.

collision is the superposition of circular motion about the center-of-mass (CM) of the system conserving angular momentum $\vec{L}$, and translation of the CM conserving momentum $\vec{p}$.

Draw a clear diagram with the appropriate coordinate system and solve for the motion of the system after the collision.

Problem 2: Practice in Setting up and Calculating Moments of Inertia

Calculate the moment of inertia for the following rigid objects:

1. A thin ring of mass M and radius R, around an axis through the center, perpendicular to the plane.

2. A disk of mass M and radius R, around an axis through the center, perpendicular to the plane.

3. A cone of mass M, height h, and radius at its base R, around the the axis of symmetry.

4. A thin rod of mass M and length L, around an axis through the center, perpendicular to the rod.

5. A thin rod of mass M and length L, around an axis through one end, perpendicular to the rod.

6. Extra Credit: A disk of mass M and radius R, around an axis in the plane of the disk and through its center.

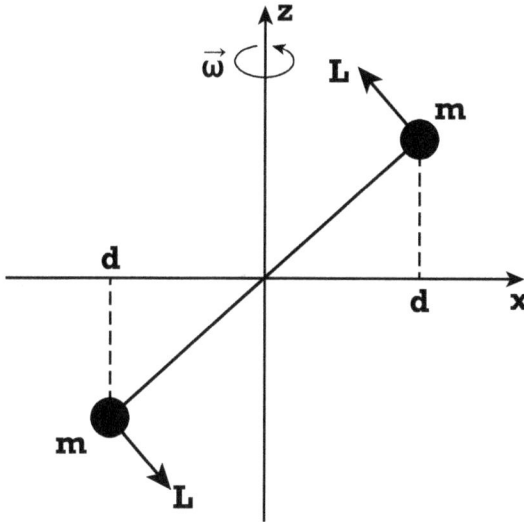

Figure 8.8. The Tinkertoy rotator.

Problem 3: Moments of an Asymmetric Rigid Rotator

Consider the tilted Tinkertoy rotator shown in Figure 8.8.

1. Calculate the moment of inertia tensor $\overset{\leftrightarrow}{I}$ in terms of the masses and the length of the rods.

2. The rotator is spun about the vertical axis at 4 Hz. Take the length of the rods to be 20 cm and each mass to be 100 gm. Find the vectors for $\vec{L}$ and $\vec{\omega}$.

3. Does $\vec{L}$ lie along $\vec{\omega}$? Why or why not? (Be as succinct (quantitative!) as you can.)

Problem 4: The Inertia Tensor $\overset{\leftrightarrow}{I}$

1. Explicitly show that the diagonal and off-diagonal elements of the moment of inertia tensor are given by:

$$I_{ij} = \int \int \int \rho(\vec{r}) d^3r \, [\delta_{ij} x_k x_k - x_i x_j] \qquad (8.27)$$

2. Can you construct a rigid body for which all the diagonal moments of inertia are zero? If so, give an example; if not, explain why not.

Problem 5: Moments and Kinetic Energy

Two cans of soup are rolled down a ramp without slipping. One can is bouillon (very watery) and one is black bean soup (essentially solid). What are the velocities of the cans at the bottom of the ramp? Take the ramp to be 0.50 meters high and 2 meters long, and use the simplest possible assumptions for the soups.[13]

Problem 6: Bowling, Torque, and Circular Motion

A bowling ball is released tangential to the floor such that it initially slides smoothly along the floor with no rotation. Assume that the frictional force between the ball and the floor depends only on the normal force and is independent of velocity while the ball is sliding. Also assume the ball is a perfect uniform sphere.[14]

1. Make a diagram of your frame with an origin and labeled axis.

2. Make a Free-Body drawing showing the forces on the ball.

3. Find the velocity of the ball versus time, $v(t)$.

4. Calculate the torques on the ball.

5. Find the angular velocity of the ball versus time, $\omega(t)$.

6. The ball will roll when the contact point has zero velocity relative to the bowling alley floor. Find the time t when this happens.

Extra Credit. Problem 7: Vector (Cross) and Scalar (Dot) Products in Tensor Notation

Prove the following tensor identity by writing out the terms for $i = 1$:

$$\epsilon_{ijk}\epsilon_{ilm} = \delta_{jl}\delta_{km} - \delta_{jm}\delta_{kl} \tag{8.28}$$

where ϵ_{ijk} is the Levi-Civita tensor and δ_{ij} is the Kronecker delta (see Section A.1.8 of Appendix A).

Extra Credit. Problem 8: An Elegant Proof of the BAC-CAB Rule

Use Equation 8.28 to prove the BAC-CAB rule for triple products.[15]

$$\vec{A} \times (\vec{B} \times \vec{C}) = \vec{B}(\vec{A} \cdot \vec{C}) - \vec{C}(\vec{A} \cdot \vec{B}). \tag{8.29}$$

---

[13] Damien, you corrected me in public on this problem when you were in the 4th grade. I hope you are doing well where-ever you are!

[14] I thank Paul Rubinov for pointing out this really isn't so for a ten-pin bowling ball. You can look it up.

[15] Simpler than it looks: start by writing out all the terms.

CHAPTER 9

Central-Force Motion

9.1 Introduction

The motion of the sun, the moon, the planets, and the night sky has been a central theme in theology and literature of many cultures from ancient times. In modern times exploring our moon, our neighbor planets, and the study of exoplanets and the search for life on them expands our horizons, literally, beyond our existence here on Earth.[1]

The history, as Western physicists tell it [17], is fascinating if complex. In 1543 Copernicus proposed the Earth moved around the Sun [18]. As a wonderful example of the role of instrumentation in major shifts of understanding,[2] the Danish astronomer Tycho Brahe invented and built instruments to measure planetary positions at an unprecedented level of precision [19]. Johannes Kepler then acquired Brahe's log books[3] and developed the basis of the modern analysis of a two-body system under a central inverse-square force, now called "The Kepler Problem" [19]. The motions of the Heavens are not due to crystalline spheres [18], but instead have been understood through high-quality data and an underlying mathematics [3, 20, 21, 22].

This chapter differs pedagogically from the previous chapters. Here the mathematics is new and difficult—solving for the equations for an elliptical orbit requires solving an integral equation. We consequently focus on developing a largely qualitative understanding of the behavior of planets.[4] A more detailed treatment is available in a number of the excellent intermediate and advanced texts in the Recommended Reading list.

The organization of the chapter is as follows. In Section 9.2 we define the problem of the Earth orbiting the Sun in terms of both bodies orbiting the center-of-mass

---

[1] Part II opens with a remarkable photograph of our own multiple orbiting systems.

[2] Both new instrumentation *and* new mathematics allow one to see previously undetected aspects of the Universe [21].

[3] Possibly by murdering him [19].

[4] Also this chapter appears at the end of the course, when there isn't enough time remaining, especially if the course is one-quarter rather than a semester, to have the conversations and group problem solving integral to collaborative learning. It's different.

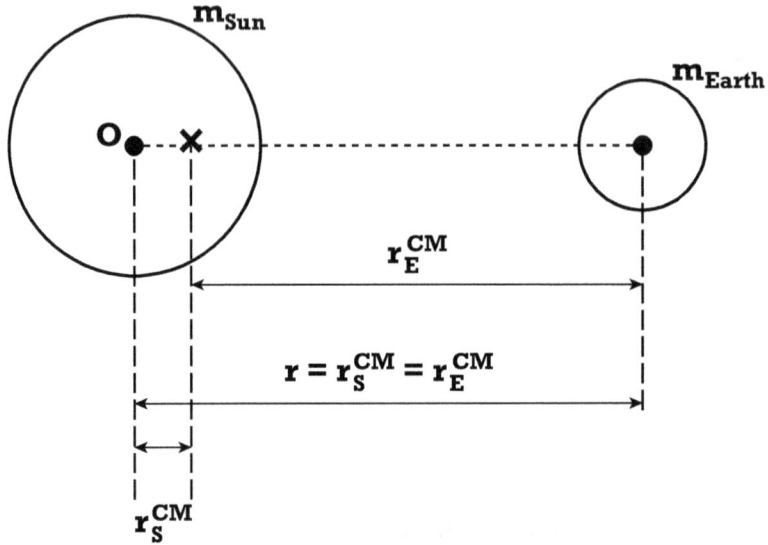

Figure 9.1. The Sun and the Earth, approximated as perfect spheres, form a 2-body system bound by the central force of gravity. The cross indicates the center-of-mass (CM) of the system. We will see that the radii r_S and r_E to the CM are time-dependent, with the shapes of the orbits being determined by the energy and angular momentum of the system, E and L.

(CM) of the system. In Section 9.3 we show that the two-body system of the Earth orbiting the Sun can be reduced to that of a single body of "reduced mass" orbiting the CM of the two-body system. In Section 9.4 we present the orbit equation that predicts the distance to the CM as a function of the angular position of the body and the constants of the motion,[5] total energy E, and angular momentum L, and show that the resulting orbits are ellipses (Kepler's First Law). We also relate the minimum (perihelion) and maximum (aphelion) radii to the body from the Sun (Helios in Greek mythology) to the orbit parameters.

Section 9.5 discusses kinetic and potential energy of the system, and the role of angular momentum conservation. Section 9.6 gives a pictorial argument for the dependence of velocity on radius, including the minimum velocity at aphelion (farthest from the Sun) and maximum velocity at perihelion (closest to the Sun). The chapter ends with Kepler's Three Laws in Section 9.7 [18, 19].

9.2   The Central Force Two-Body Problem

Figure 9.1 shows the Sun and the Earth. Both orbit the CM of the two-body gravitationally-bound system at radii r_S and r_E respectively. Note that the radii change with time according to the angular position in the orbit.

---

[5] The phrase is equivalent to "quantities conserved in the motion."

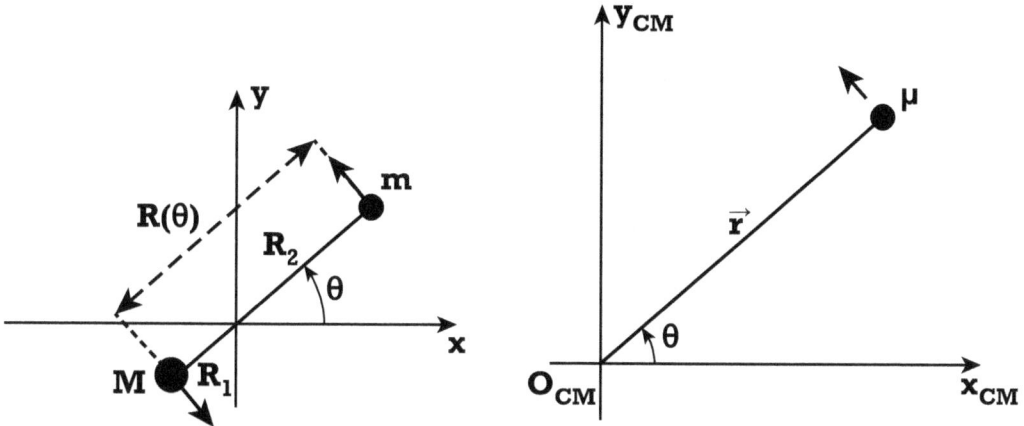

Figure 9.2. Left: The two-body system rotating in the xy plane about the CM. Right: The equivalent one-body system of reduced mass μ. Note that at the same θ the radius r of the orbit of the single body about the CM is equal to the original distance R between the Sun and Earth; the restatement from a 2-body to a 1-body problem is absorbed entirely in the orbiting reduced mass. The angular frequency is also unchanged.

The figure shows the system in a coordinate system with the origin at the center of the Sun, and the positive x-axis to the right. The cross indicates the center-of-mass (CM) of the system. The motion corresponds to the Sun and the Earth co-rotating about the CM of the system at angular velocity ω.

9.3   The Equivalent One-Body Problem

The Kepler problem of two bodies each orbiting the center-of-mass can be reduced to the problem of a single object with "reduced mass" orbiting the center-of-mass (CM) as shown in Figure 9.2. This can be solved for the orbit equation $r(\theta)$, which gives the distance of the object to the CM as a function of the angular position on the orbit, and the radial equation, $r(t)$, which gives the distance versus time.[6]

9.3.1   Calculating the Reduced Mass μ

We define the mass of the Sun and the mass of the Earth to be m_S and m_E, respectively. They are separated by a distance r that depends on the angle θ measured from the origin.

---

[6] Note the radial distance $r(t)$ of the orbit is unchanged in going from the 2-body to the 1-body problem, i.e., the one-body $r(t)$ is the distance between the Sun and Earth for that angular position; the restatement from a 2-body to a 1-body problem is entirely made by using the orbiting reduced mass. The angular frequency is also unchanged.

The calculation of the reduced mass proceeds in the following steps:

1. **In the Sun's frame express r_S and r_E in terms of m_S, m_E, and r.**
   The radial distance from the origin to the CM is given by:

$$r_{CM} = \frac{\sum_i m_i r_i}{\sum_i m_i}$$

(9.1)

$$= \frac{m_E r_E + m_S r_S}{m_E + m_S}.$$

In the Sun's frame:

$$r_E = r$$
$$r_S = 0$$

(9.2)

$$r_{CM} = \frac{m_E}{m_E + m_S} r.$$

2. **Transform to the CM frame where $r_{CM} = 0$.**
   In the CM frame:

$$r_E = r - r_{CM}$$
$$= \frac{m_S}{m_S + m_E} r$$
$$r_S = -r_{CM}$$
$$= -\frac{m_E}{m_S + m_E} r$$

(9.3)

and the distance r between the Sun and the Earth is unchanged by the translation of the origin (of course).

3. **Express the Moment of Inertia I of the Two-Body System in Terms of a Single Body of Mass μ (mu).**
   The moment of inertia of the two-body system about the CM is the sum of the masses times their respective radii squared:

$$I_{Tot} = m_S r_S^2 + m_E r_E^2$$
$$= [m_S(\frac{m_E}{(m_S + m_E)})^2 + m_E(\frac{m_S}{(m_S + m_E)})^2] r^2$$
$$= [\frac{m_S m_E (m_S + m_E)}{(m_S + m_E)^2}] r^2$$
$$I_{Tot} = \frac{m_S m_E}{m_S + m_E} r^2.$$

(9.4)

This is the moment of inertia of a single body of mass μ orbiting the center of mass at radius r, where the reduced mass μ is:

$$\mu = \frac{m_S m_E}{m_S + m_E}.$$ (9.5)

As an aside, this form, the product of 2 quantities over the sum of the same 2 quantities, is always a tip-off that the quantities add in the inverse, e.g.:

$$\frac{1}{\mu} = \frac{1}{m_S} + \frac{1}{m_E}.$$ (9.6)

Having transformed the two-body problem into a one-body problem, we can abandon using the Greek letter μ and instead just use m for the orbiting reduced mass.

9.4   Elliptical Orbits: The Orbit Equation $r(\theta)$

The orbit equation gives the distance of an orbiting object from the CM, $r(\theta)$, as a function of the angular position in the orbit, θ. For the gravitational potential $V(r) = -k/r$, solving for the orbit equation is not easy.[7] The orbit is completely determined by four constants: the gravitational coupling k, the (reduced) mass m, and the two kinematic "constants of the motion" (conserved quantities), the energy E, and the angular moment L.[8]

The orbit equation can be written [23] in the form:

$$r(\theta) = \frac{\alpha}{1 + \epsilon \cos \theta}$$ (9.7)

which is the equation for an ellipse.[9] The parameters α and ϵ are two new constants constructed from k, m, E, and L.[10]

9.4.1   The Two Ellipse Parameters a and b

Ellipses are usually characterized by the semi-major axis a and the semi-minor axis b, as shown in Figure 9.3. These two parameters capture the orbit's shape, but not

---

[7] The details of the derivation can be skipped with no loss, but for the curious and brave, are well described in Goldstein, for example [23].

[8] Note that the convention is that E is negative for bound states; $E = 0$ is the threshold for escape from the gravitational potential well, and $E > 0$ means it's not coming back.

[9] From Brahe's data, Kepler recognized the orbits as ellipses. Gingerich points out Kepler's accurate determination of the position of the focus of the ellipse from Brahe's data was essential to the identification of the shape, and is little discussed [18].

[10] Note the degeneracy: the orbit equation uses only the two constants α and ϵ to describe the elliptical orbit, while there are 4 physical parameters. Orbits form only for the values of k, m, E, and L, and hence α and ϵ, that satisfy Eq. 9.7.

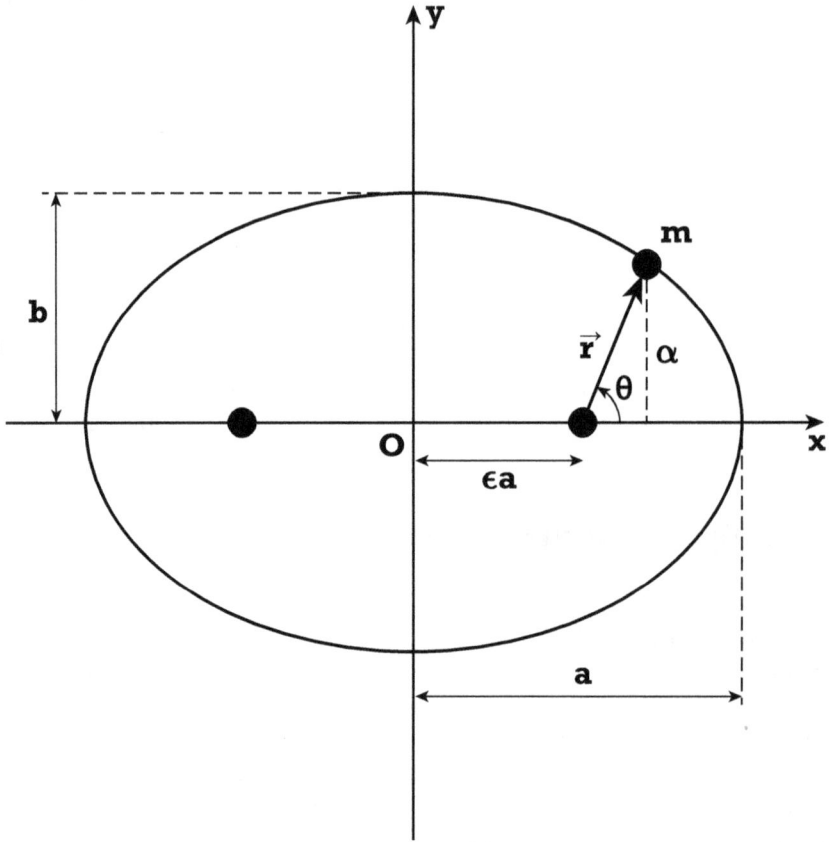

Figure 9.3. An elliptical orbit with semi-major axis a and semi-minor axis b. The position of the (reduced) mass m in the orbit is parameterized by the distance to a focus $r(\theta)$ and the angle θ measured there to the x-axis. An alternative pair of parameters is shown: the distance between the center of the ellipse and the focus, ϵa, where ϵ is the dimensionless *eccentricity*; and α is the *semi-latus rectum*, the distance normal to the major axis from the focus to the orbit. Note that when the eccentricity $\epsilon = 0$, the focus and the origin coincide, and the orbit is a circle.

the position of the planet as it orbits. The dependence of a and b on the constants of the motion E, K, and L, is the subject of Problem 5 in this Chapter.

9.4.2   The Two Ellipse Parameters α and ϵ

The parameter α, called the *semi-latus rectum* has dimensions of length, and characterizes the size of the ellipse, as shown in Figure 9.3. It depends on the angular momentum, mass, and the strength of the gravitational force k as the fraction:

$$\alpha = \frac{L^2}{mk}. \tag{9.8}$$

Interestingly, for a given reduced mass and coupling of the force k, α depends only on the orbital angular momentum L, and **not** the energy E.

The second parameter is ϵ, the *eccentricity*, a dimensionless quantity that characterizes the deviation of the orbit from a circle. For given values of m and k the eccentricity depends on both the energy[11] and the angular momentum:

$$\epsilon = \sqrt{1 + \frac{2\,E\,L^2}{m\,k^2}}. \tag{9.9}$$

Figure 9.3 shows the geometry: ϵ is the fraction of the semi-major axis between the focus and the center of the ellipse. A circle has eccentricity $\epsilon = 0$.

9.4.3   Finding the Minimum (Perihelion) and Maximum (Aphelion) Radii of the Orbit

Equation 9.7 allows expressing the semi-major and semi-minor axes of the ellipse in terms of the eccentricity ϵ and the semi-latus rectum.

The minimum distance r_{min} occurs at $\theta = 0$, the turning point of the orbit closest to the Sun when the planet is moving fastest:

$$r(\theta = 0) = \frac{\alpha}{1 + \epsilon \cos(0)}$$
$$r(0) = \frac{\alpha}{1 + \epsilon}. \tag{9.10}$$

Similarly the maximum distance r_{max} occurs at $\theta = \pi$, the turning point of the orbit farthest to the Sun when the planet is moving slowest:

$$r(\theta = \pi) = \frac{\alpha}{1 + \epsilon \cos(\pi)}$$
$$r(\pi) = \frac{\alpha}{1 - \epsilon}. \tag{9.11}$$

9.4.4   Relating the Semi-Latus Rectum and the Semi-Major Axis

The sum of r_{min} and r_{max} equals the length of the major axis:

$$r_{min} + r_{max} = 2a. \tag{9.12}$$

Substituting in the expressions for r_{min} and r_{max}:

$$\frac{\alpha}{1 + \epsilon} + \frac{\alpha}{1 - \epsilon} = 2a$$

---

[11] Be careful! Remember that the energy E for bound states is negative. The eccentricity is less than 1 for an ellipse.

$$\frac{\alpha(1-\epsilon)+\alpha(1+\epsilon)}{1-\epsilon^2} = 2a \tag{9.13}$$

$$\frac{2\alpha}{1-\epsilon^2} = 2a.$$

Solving for α:

$$\alpha = (1-\epsilon^2)a. \tag{9.14}$$

The semi-latus rectum α is proportional to the semi-major axis a. When the eccentricity is zero the orbit is a circle, and both α and a are equal to the radius R.

9.5   Motion of a Mass of Energy E and Angular Momentum L in a Potential Well

Our planet Earth is (happily) captured by the Sun; it would take adding kinetic energy to make it free. With the convention that the potential $V(r) = -k/r$ (i.e., no constant term), freedom occurs at total energy $E = 0$, and the kinetic energy T has only to be infinitesimally positive for escape. However, since the potential is negative and goes to $-\infty$ as $r \to 0$, as shown by the dotted line in Figure 9.4, one can ask what has kept our planet from catastrophically sliding down the potential to $r = 0$? By exploiting the constants of motion E and L, we will derive the answer.

9.5.1   The Angular Momentum Barrier

The angular momentum $L = mr^2\dot{\theta}$ and the total energy E are the two kinematic constants of the motion. Conservation of L requires that as the radius r decreases, the angular velocity $v_\theta = r\dot{\theta}$ increases.[12] Conservation of energy requires there must exist a minimum radius of the orbit.[13] The effect is called the *angular momentum barrier*, as it is an answer to the question as to why we don't fall into the Sun.

9.5.2   The Effective Potential: Substituting $L^2/2mr^2$ for the Angular Kinetic Energy $T_{\dot{\theta}} = \frac{1}{2}m(r\dot{\theta})^2$

Developing an intuition for the two-dimensional motion in the plane of rotation is difficult. We consequently turn the problem into a one-dimensional motion in a modified potential. We can eliminate the kinetic energy associated with the angular velocity by parameterizing it with the (constant) angular momentum L and the radius r. The result is an equation of motion in the position r and the radial

---

[12] This is often illustrated by a spinning figure skater pulling in their arms.
[13] Note that the minimum and maximum radii occur on the major axis of the orbit by symmetry.

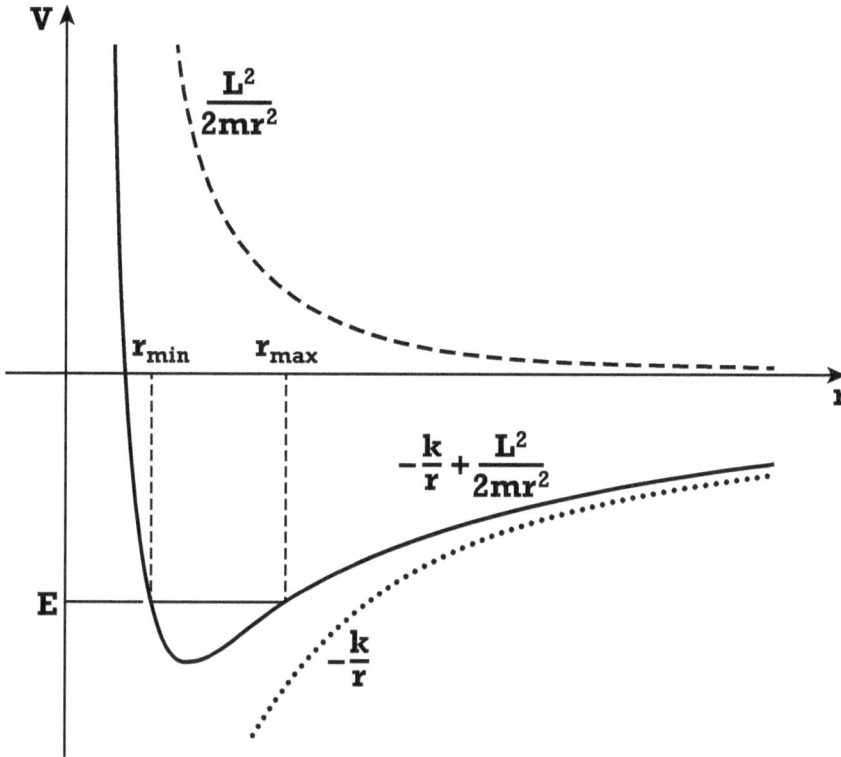

Figure 9.4. The contributions to the effective potential versus radius. The gravitational potential $V(r) = -k/r$ is shown as a dotted line. The angular momentum barrier, $L^2/2mr^2$ is shown as a dashed line. The solid line represents the sum, the "effective potential" V_{eff}. Bound states have negative energy, and consequently have radii $r_{min} \leq r \leq r_{max}$, as shown by the solid horizontal line at energy E, where $E < 0$. As the Earth orbits, the radius oscillates back-and-forth along this line at constant E between the limits r_{min} (perihelion) and r_{max} (aphelion).

velocity $\dot{r}$. We can then absorb the dependence on r into the new potential, retaining the dependence on the radial velocity $\dot{r}$ as the only term in the kinetic energy.

The mathematics is not hard. The total energy of the system is given by $E = T + V$, where T is the kinetic energy in polar coordinates,

$$T = \frac{1}{2m}(\dot{r}^2 + (r\dot{\theta})^2). \tag{9.15}$$

The angular momentum is given by

$$L = mr^2\dot{\theta}. \tag{9.16}$$

Substituting $\dot{\theta} = L/mr^2$ from Eq. 9.16 into Eq. 9.15 gives

$$T = \frac{1}{2m}\dot{r}^2 + (L^2/2mr^2) \tag{9.17}$$

and

$$E = \frac{1}{2m}\dot{r}^2 + V(r) + (L^2/2mr^2). \tag{9.18}$$

Isolating the terms that depend on position only, we define the effective potential:

$$V_{eff}(r) \equiv V(r) + L^2/2mr^2 \tag{9.19}$$

and the total energy is that of an object in a one-dimensional well with potential $V_{eff}(r)$:

$$E = \frac{1}{2m}\dot{r}^2 + V_{eff}(r). \tag{9.20}$$

Figure 9.4 shows the potential $-k/r$ as a line of short dots, the angular momentum barrier $+L^2/2mr^2$ with long dashes, and the sum, $V_{eff}(r)$, as a solid line.

Bound states have negative energy, but cannot have total energy less than the potential energy as the kinetic energy T cannot be negative. Bound states consequently have radii $r_{min} \leq r \leq r_{max}$, as shown with the solid horizontal line at energy E, where $E < 0$. States with $E > 0$ are called *free states* and are free to propagate to all values of radius.

9.5.3   Visualizing the Orbit Dependence on the Physical Parameters E, L, and k

For those of us who are visual learners, it may be helpful to think of the orbiting object oscillating in one-dimension back-and-forth between r_{min} and r_{max} along the horizontal line at energy E in Figure 9.5. In this region the kinetic energy $T = \frac{1}{2}mv^2$ is the vertical distance from the horizontal line (constant energy E) and the effective potential (V), i.e., $T = E - V$, which has a very different behavior as the planet approaches r_{max} than r_{min}.

At energy E_0, the system sits at the minimum, at constant radius R. The orbit is a circle. At energies E_1 and E_2, the radius oscillates between the respective values of r_{min} and r_{max}. The maximum distance r_{max} is determined by the strength k of the potential (solid line). The minimum distance r_{min} is determined by the steep angular momentum barrier $L^2/2mr^2$ (dashed line) , and so is seen to vary less with E than does r_{max}. The maximum distance r_{max} is determined by the gentler upward slope of the potential at large r, and changes more rapidly with energy.

9.6   Visualizing the Radial Behavior: Velocities at Aphelion, Perihelion

There are topics, such as Special Relativity and the motion of asymmetric rigid bodies covered in the first chapters of the course that I find much simpler

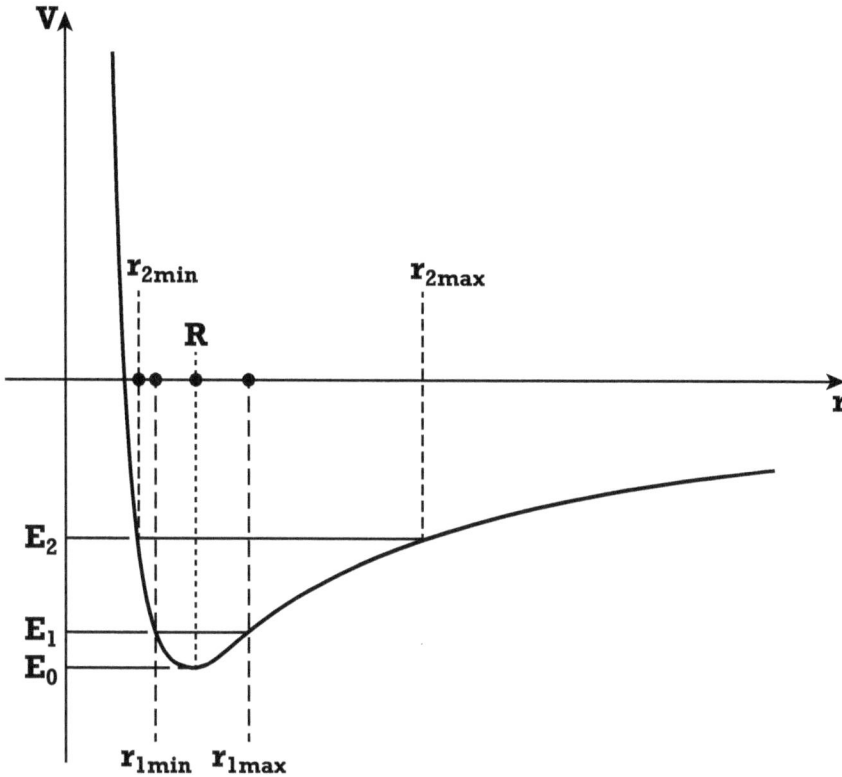

Figure 9.5. The effective potential V_{eff} plotted versus radius with the allowed radial regions for three energies, E_0, E_1 and E_2. The value E_0 is at the minimum of the effective potential, and the orbit is a circle with the radius of the minimum. For both E_1 and E_2, as the body orbits, the radius will shuttle back-and-forth at constant energy between the perihelion r_{min} and the aphelion r_{max}. The radial kinetic energy at radius r, and hence the radial velocity $\dot{r}$, is determined by $T = \frac{1}{2}m\dot{r}^2 = (E - V_{eff}(r))$; one can read the difference off the plot. The slope of V_{eff} at radius r_{min} is significantly greater than that at r_{max}, a measure of the eccentricity of the orbit and the local velocity.

algebraically[14] than visually/geometrically [24]. However, there are several problems that seem difficult algebraically and for which a visual or graphic description seems more effective pedagogically. The dependence of the shape of the elliptical orbit on the two constants of the motion E and L and the force constant k is one such.

Figure 9.5 again shows the effective potential well in which the planet sits, but for three energies, E_0, E_1, E_2. Bound states have negative energy, and consequently have radii $r_{min} \leq r \leq r_{max}$, as indicated at the intersections of the horizontal lines of constant E with the effective potential.

At energy E_0, the system sits at the minimum of the effective potential (solid line), at constant radius R, i.e., $r_{min} = r_{max}$. The orbit is a circle, with $\epsilon = 0$ and

---

[14] Both were treated using simple matrix multiplication.

$\alpha = a$. At energies E_1 and E_2, the system oscillates between the respective values of r_{min} and r_{max}. The ellipses have successively larger values of eccentricity as E approaches zero from below.

9.7   Kepler's Three Laws

Tycho Brahe (1546–1601) was a Danish nobleman and astronomer who was given the island of Hven [18, 19]. He was much hated by the local farmers for conscripting them to build his observatory,[15] but enormously talented as an instrument builder and precise observer of planetary motion. From Brahe's logbooks Kepler (1571–1630) distilled the orbital motion of the moons of Jupiter into three Laws [19]. Motion under an inverse-square attractive central force is still known as "The Kepler Problem."

9.7.1   Kepler's First Law

Kepler's First Law:

Planetary Orbits are ellipses with the Sun at one focus.

Figure 9.3 shows the orbit of the reduced mass, with the origin being the center of mass of the system (very close to the center of the Sun). The orbit equation represents an ellipse; $r(\theta)$ is the distance to one focus:

$$r(\theta) = \frac{\alpha}{1 + \epsilon \cos(\theta)} \tag{9.21}$$

where ϵ is the eccentricity; and α is the semi-latus rectum.

9.7.2   Kepler's Second Law

The planet sweeps out equal areas in equal times.

This is a statement of conservation of angular momentum,

$$L = mr^2 \dot{\theta} = mr(r\dot{\theta}). \tag{9.22}$$

Figure 9.6 shows two triangles swept out in different places in the orbit; the area is proportional to the radius r times the distance traveled in time Δt:

$$\Delta A = (r)(r\dot{\theta})\Delta T$$
$$\Delta A / \Delta T = r^2 \dot{\theta} \tag{9.23}$$
$$= L/m$$

which is constant.

---

[15] Personal communication from a tour guide on the island of Hven.

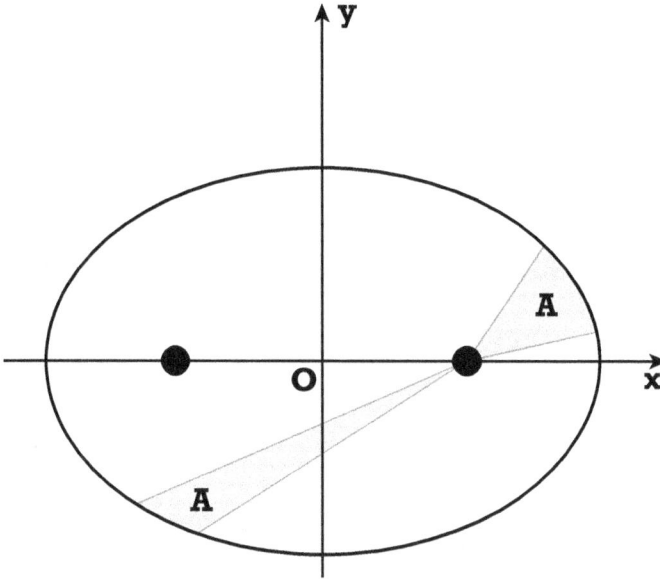

Figure 9.6. Kepler's Second Law: The planet sweeps out equal areas in equal times. The planet moves faster at smaller r, making the angle $\Delta\theta$ larger than at large r. The "equal areas in equal times" law is a visual statement that the angular momentum $L = mr^2\dot{\theta}$ is constant.

9.7.3   Kepler's Third Law

The square of the period of the orbit is proportional to the semi-major axis cubed: $\tau^2 \propto a^3$.

We prove this for the simplest case of a circular orbit of radius R, for which the period τ is equal to the circumference divided by the velocity:

$$\tau = \frac{2\pi R}{v}$$
$$\tau^2 = \frac{4\pi^2 R^2}{v^2}. \tag{9.24}$$

From Newton's Second Law

$$ma = F; \quad a = -\frac{v^2}{R}; \quad F = -\frac{k}{R^2}$$
$$m(\frac{v^2}{R}) = \frac{k}{R^2} \implies v^2 = \frac{k}{mR}. \tag{9.25}$$

Substituting v^2 into Eq. 9.24:

$$\tau^2 = 4\pi^2 R^2 (\frac{mR}{k}) = 4\pi^2 (\frac{m}{k})R^3 \tag{9.26}$$

as stated in the Third Law.

Two factors of R come from the period-squared (τ^2); one additional factor comes from the inverse-square force equation.

9.8   Problem Set 9: Central Force Motion, Kepler's Laws

Study Groups: These problems are time-consuming and non-intuitive at first. To finish efficiently, work them with your study group. However, the work you hand in **has to be your own**.

Problem 1: Reduced Mass

It pays to check a derived formula with limiting cases. Calculate the limit of the reduced mass μ for masses m_1 and m_2 in the following common cases:

1. $m_1 \gg m_2$

2. $m_2 \gg m_1$

3. $m_1 = m_2 \equiv m$.

Problem 2: Reduced Mass in the (Our!) Solar System

1. Derive the formula for the reduced mass for two gravitationally-bound objects (Equation 9.6).

2. Calculate the reduced mass μ_{ME} for the Moon and the Earth. What fraction of the Moon's mass is the reduced mass?

3. Calculate the displacement between the CM and the center of the Earth for the Moon and the Earth (assume what you need to).

4. Calculate the reduced mass μ_{ES} for the Earth and the Sun. What fraction of the Earth's mass is the reduced mass?

5. Calculate the displacement between the CM and the center of the Sun for the Earth and the Sun.

Problem 3: Kepler's First Law: The Planetary Orbits are Ellipses

For an elliptical orbit, the orbit is given in terms of (r, θ) by:

$$r = \frac{\alpha}{(1 + \epsilon cos(\theta))} \tag{9.27}$$

where α is the *semi-latus rectum*[16] and ϵ is the *eccentricity*.

---

[16] I kid you not.

1. Find the semi-major axis a in terms of α and ϵ.

2. Find the semi-minor axis b in terms of α and ϵ.

3. Find the area in terms of α and ϵ.

4. Starting with the equation for an ellipse:

$$(x/a)^2 + (y/b)^2 = 1 \tag{9.28}$$

find the area in terms of a and b by mapping the ellipse to the unit circle (i.e., change variables so that the equation is that of a circle.).

Problem 4: Kepler's Second Law

State and prove Kepler's Second Law.

Problem 5: Two Other Parameters That Specify an Ellipse

Two parameters that specify a planetary orbit are the semi-major latus rectum and the eccentricity:

$$\alpha = \frac{L^2}{\mu k} \qquad \epsilon = \sqrt{1 + \frac{2EL^2}{\mu k^2}}. \tag{9.29}$$

There is another pair of parameters that specify the orbit, the semi-major and semi-minor axes a and b. Express a and b in terms of μ, E, k, and L. Explicitly point out the dependence of a and b on E and L.

Problem 6: Kepler's Third Law

State and prove Kepler's Third Law for circular orbits.

Problem 7: Our Local Neighborhood

For the eight planets, make a table of the semi-major and semi-minor axes, eccentricity, and period. Calculate τ^2/a^3 using years and a.u. as units.

Problem 8: The Exile of Pluto

Do the same as in Problem 7 for Pluto. Make a plot (histogram) for a, b, ϵ and τ for all the planets and Pluto, with labels identifying each of them (you can do these by hand and photograph them if you wish—no need to be fancy, only clear and neat).

APPENDIX

The mathematical methods in Appendix A form an essential parallel ingredient of the text. They provide the tools and language crucially the notation used for each topic is the when they are needed to learn the physics. The methods and concepts as well as the vocabulary are the framework for the discussions in the Study Group solving the Problem Sets, and so subtle concepts get clarified by being discussed continuously.[1]

It has become a common practice to out-source the teaching of mathematical methods for introductory physics courses to the Math Department. This structure has a number of weak points.[2] The methods are almost always introduced in a different time and completely different context from the lecture on the physics. The notation is usually different, often in ways that lead to serious confusion. The setting of the syllabus and sequence of topics must respond to the needs of departments other than physics such as chemistry and molecular engineering, which can lead to large-scale synchronization problems, such as the ordering of linear algebra versus multi-variable Partial differential equations. The large number of students from multiple departments also requires many parallel sections, not all of which are taught by highly experienced instructors in good communication with the Physics Department.

But most importantly, the service courses short-circuit many physics majors from seeking out the most gifted teachers and most interesting courses in the Math Department, a remarkable opportunity that is lost by taking one of the terminal service sequences.

The Appendix is an attempt at providing the mathematical methods needed to solve the Problem Sets as the course proceeds. All errors and misconceptions are my own; I recommend having a critical eye, and I welcome suggestions for improvements from students, graduate TAs, and instructors.

---

[1] The process is (of course) the same as learning a foreign language; immersion in a non-English-speaking environment results in a much deeper and more comfortable understanding.

[2] I have watched many attempts for more than 50 years; the structure has never consistently worked well, and I will bet money that it never will.

A.1   Mathematical Methods and Conventions

A.1.1   Vector and Matrix Multiplication

We will do examples in 3-dimensional Cartesian space; however, the technique is the same in 2-dimensions.

A.1.1.1   Vector on Vector: the Dot Product (Scalar Product, Projection Operator)

We start with an example of the scalar product of two vectors,

$$\vec{A} \cdot \vec{B} = A_x B_x + A_y B_y + A_z B_z. \tag{A.1}$$

In index notation,

$$\vec{A} \cdot \vec{B} = \sum_{i=1,3} A_i B_i. \tag{A.2}$$

In index notation with Einstein summation:

$$\vec{A} \cdot \vec{B} = A_i B_i. \tag{A.3}$$

A.1.1.2   Vector on Vector: the Cross Product (Vector Product)

The cross product of vectors $\vec{A}$ and $\vec{B}$ is a vector itself. In explicit Cartesian notation, the components are given by:

$$(\vec{A} \times \vec{B})_x = A_y B_z - A_z B_y$$
$$(\vec{A} \times \vec{B})_y = A_z B_x - A_x B_z \tag{A.4}$$
$$(\vec{A} \times \vec{B})_z = A_x B_y - A_y B_x.$$

In index notation:

$$(\vec{A} \times \vec{B})_i = A_j B_k - A_k B_j \tag{A.5}$$

where i, j, k cycle through 1,2,3 and back to 1 (cyclic).

In index notation with Einstein summation:

$$\vec{A} \times \vec{B}_i = \epsilon_{ijk} A_j B_k. \tag{A.6}$$

A.1.1.3   The Right-Hand Rule for the Cross Product

The Cartesian coordinate system comprising the 3 orthogonal axes in the $(\hat{x}, \hat{y}, \hat{z})$ directions has an ambiguity: $\hat{x} \times \hat{y}$ can be $\pm \hat{z}$. The positive sign produces a Right-Handed System; the negative sign a Left-Handed System. We chose the former as our convention.

The Right-Hand Rule is: Place the heel of your right hand on the origin and extend your fingers along $\vec{A}$ with your thumb perpendicular to the plane formed by $\vec{A}$ and $\vec{B}$. Now bend your fingers in the direction of $\vec{B}$. Your thumb now points in the direction of $\vec{A} \times \vec{B}$. It may or may not speak to you; I find index notation much quicker.[3]

A.1.1.4   *Matrix on Vector: Transforming a Vector*

Multiplying a vector by a matrix consists of the same operation of multiplying a vector by a vector (Section A.1.1.1), but doing it 3 times, once for each row of the multiplying matrix.

Consider the example of matrix $\overset{\leftrightarrow}{M}$ operating on vector $\vec{B}$ to produce a new vector $\vec{A}$:

In vector notation:
$$\vec{A} = \overset{\leftrightarrow}{M}\vec{B}. \tag{A.7}$$

In matrix notation
$$\begin{pmatrix} A_x \\ A_y \\ A_z \end{pmatrix} = \begin{pmatrix} M_{xx} & M_{xy} & M_{xz} \\ M_{yx} & M_{yy} & M_{yz} \\ M_{zx} & M_{zy} & M_{zz} \end{pmatrix} \begin{pmatrix} B_x \\ B_y \\ B_z \end{pmatrix}. \tag{A.8}$$

Note that reading the equation for A_x as the scalar product of the top row of $\overset{\leftrightarrow}{M}$ treated as a vector with the vector $\vec{B}$ gives:

$$\begin{pmatrix} A_x \\ \end{pmatrix} = \begin{pmatrix} M_{xx} & M_{xy} & M_{xz} \end{pmatrix} \begin{pmatrix} B_x \\ B_y \\ B_z \end{pmatrix} \tag{A.9}$$

Similarly taking the dot product of the 2nd row of $\overset{\leftrightarrow}{M}$ on $\vec{B}$ gives:

$$\begin{pmatrix} A_y \\ \end{pmatrix} = \begin{pmatrix} M_{yx} & M_{yy} & M_{yz} \end{pmatrix} \begin{pmatrix} B_x \\ B_y \\ B_z \end{pmatrix}. \tag{A.10}$$

and analogously for A_z as the dot product of the 3rd row of $\overset{\leftrightarrow}{M}$.

Note that the second index on each matrix element corresponds to the component of the vector being transformed by $\overset{\leftrightarrow}{M}$. The first index corresponds to the component of the transformed vector. The matrix comprises the coefficients of the three linear equations for each of the components of $\vec{A}$.

A.1.1.5   *Matrix on Matrix: Product of Two Transformations*

Multiplying a matrix by a matrix consists of the same operation of multiplying a matrix on a vector (Section A.1.1.4), but doing it 3 times, once for each column of the matrix being transformed.

---

[3] Look Ma! No Hands!

$$\begin{pmatrix} A_{11} & A_{12} & A_{13} \\ A_{21} & A_{22} & A_{23} \\ A_{31} & A_{32} & A_{33} \end{pmatrix} = \begin{pmatrix} B_{11} & B_{12} & B_{13} \\ B_{21} & B_{22} & B_{23} \\ B_{31} & B_{32} & B_{33} \end{pmatrix} \begin{pmatrix} C_{11} & C_{12} & C_{13} \\ C_{21} & C_{22} & C_{23} \\ C_{31} & C_{32} & C_{33} \end{pmatrix}. \tag{A.11}$$

This is, to find A_{11} take the scalar product of the first row of $\overset{\leftrightarrow}{B}$ and the first column of $\overset{\leftrightarrow}{C}$; to find A_{12} take the scalar product of the first row of $\overset{\leftrightarrow}{B}$ and the second column of $\overset{\leftrightarrow}{C}$, and to find A_{31} take the scalar product of the third row of $\overset{\leftrightarrow}{B}$ and the first column of $\overset{\leftrightarrow}{C}$.

Exercise for the reader: Rotating by the same angle twice

$$\begin{pmatrix} \cos(\theta_1+\theta_2) & \sin(\theta_1+\theta_2) & 0 \\ -\sin(\theta_1+\theta_2) & \cos(\theta_1+\theta_2) & 0 \\ 0 & 0 & 1 \end{pmatrix} = \begin{pmatrix} \cos(\theta_2) & \sin(\theta_2) & 0 \\ -\sin(\theta_2) & \cos(\theta_2) & 0 \\ 0 & 0 & 1 \end{pmatrix} \begin{pmatrix} \cos(\theta_1) & \sin(\theta_1) & 0 \\ -\sin(\theta_1) & \cos(\theta_1) & 0 \\ 0 & 0 & 1 \end{pmatrix}. \tag{A.12}$$

Following the rules for matrix multiplication, Eq. A.8, yields the formulas for the sin and cos of twice the angle, $\sin 2(\theta)$ and $\cos 2(\theta)$:

$$\sin(2\theta) = 2\sin\theta\cos\theta$$
$$\cos(2\theta) = \cos^2(\theta) - \sin^2(\theta). \tag{A.13}$$

A.1.2   Total and Partial Derivatives

A partial derivative is the operation of differentiating with respect to one variable holding all other variables constant. A total derivative may include contributions from other variables which have a secondary dependence on the variable being changed via the Chain Rule of calculus.

Our convention is that a total derivative is represented by $\frac{df(x,t)}{dx}$, and a partial derivative by $\frac{\partial f(x,t)}{\partial x}$.

As an example, consider a function of two variables, $f(x,t)$. The total derivative with respect to time, depends directly on t. However, if x is time-dependent, $f(x,t)$ will change with time through the dependence on x. Using the Chain Rule:

$$\frac{df(x,t)}{dx} = \frac{\partial f(x,t)}{\partial t} + \frac{\partial f(x,t)}{\partial x}\frac{\partial x}{\partial t}. \tag{A.14}$$

A.1.3   Taylor and Maclaurin Series, and the Ubiquity of Approximation

As an experimentalist, I feel keenly that our knowledge of the physical world is only as good as measurements have shown, in contrast to the mathematics of the language we use to describe it.[4] Acquiring a robust understanding of the limits of

---

[4] For a strongly-held—but to-be-respected—opposing view, see Ref. [20].

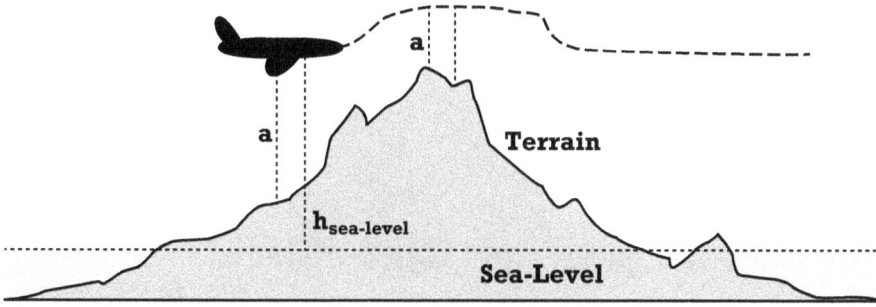

Figure A.1. An example illustrating total and partial derivatives. An airplane flies across a plain and then over a mountain range. The plane's altimeter measures the altitude, the height above the ground, as a function of the plane's position and the time, $a(x, t)$. The height above sea level, $h_{sl}(x)$ can be measured by radar. The altitude depends both on the topography under the plane's position and whether or not the plane has been climbing or descending; the height above sea level depends only on the past vertical motion and not on the position. The total derivative of the altitude is given by applying the Chain Rule (Equation A.14): $\frac{da(x,t)}{dt} = \frac{\partial a}{\partial t} + \frac{\partial a}{\partial x}\frac{\partial x}{\partial t}$, where $\frac{\partial x}{\partial t} \equiv v$, the speed of the plane.

precision and the use of approximations should be a goal of an introductory course. We routinely quantify the limits of precision by expanding exact solutions in infinite expansions and using only as many terms as we can justify. However, a word of good advice for Physics students who treasure exactitude: take as much Math as you can from your Math Department. I had wonderful math professors, and like music, it provides lifetime joy.

A.1.3.1 Taylor Series

Consider a continuous one-dimensional function f(x), for which we know the value at a location $x = a$, $f(x = a) \equiv f(a)$. The values of f(x) near a can be expressed as an infinite series in the derivatives of $f(x)$ evaluated at $x = a$:

$$f(x) = f(a) + \frac{1}{1!}\frac{\partial f}{\partial x}(x - a) + \frac{1}{2!}\frac{\partial^2 f}{\partial x^2}(x - a)^2 + \dots \frac{1}{n!}\frac{\partial^n f}{\partial x^n}(x - a)^n + \dots \quad (A.15)$$

where the derivatives $\frac{1}{n!}\frac{\partial^n f}{\partial x^n}$ are evaluated at $x = a$.

A.1.3.2 Maclaurin Series

The Maclaurin series is the special case where the function is expanded about the origin, $x = 0$:

$$f(x) = f(0) + \frac{1}{1!}\frac{\partial f}{\partial x}(x) + \frac{1}{2!}\frac{\partial^2 f}{\partial x^2}(x)^2 + \dots \frac{1}{n!}\frac{\partial^n f}{\partial x^n}(x)^n + \dots \quad (A.16)$$

where the derivatives $\frac{1}{n!}\frac{\partial^n f}{\partial x^n}$ are evaluated at $x = 0$.

A.1.3.3   The $(1+x)^\alpha \approx 1 + \alpha x$ Expansion for x << 1

There is a special case of the Taylor series that comes up often: $(1+x)^\alpha$ where $x << 1$.

$$(1+x)^\alpha = 1 + \alpha x + \frac{1}{2!}(\alpha)(\alpha - 1)x^2 + \dots \frac{1}{n!}(\alpha)\dots(\alpha - n + 1)x^n + \dots \quad \text{(A.17)}$$

A.1.3.4   The Taylor Expansions of e^{ix}, $\cos\theta$, and $\sin\theta$

The decomposition of $e^{i\theta}$ into Real and Imaginary parts is an essential tool. We derive it here and discuss it in Appendix A.1.4.

Expand $f(x) = e^{ix}$ in a Maclaurin series:

$$f(x) = f(0) + \frac{1}{1!}\frac{\partial f}{\partial x}(x) + \frac{1}{2!}\frac{\partial^2 f}{\partial x^2}(x)^2 + \frac{1}{3!}\frac{\partial^3 f}{\partial x^3}(x)^3 + \frac{1}{4!}\frac{\partial^4 f}{\partial x^4}(x)^4 + \dots \quad \text{(A.18)}$$

Taking the derivatives and evaluating at $x = 0$:

$$
\begin{aligned}
f(0) &= 1 \\[2mm]
\frac{1}{1!}\frac{\partial f}{\partial x}\Big|_{x=0} x &= +ie^{ix}\Big|_{x=0} x = ix \\[2mm]
\frac{1}{2!}\frac{\partial^2 f}{\partial x^2}\Big|_{x=0}(x)^2 &= -\frac{1}{2!}e^{ix}\Big|_{x=0} x^2 = -x^2/2 \\[2mm]
\frac{1}{3!}\frac{\partial^3 f}{\partial x^3}\Big|_{x=0}(x)^3 &= -i\frac{1}{3!}e^{ix}\Big|_{x=0} x^3 = -ix^3/3! \\[2mm]
\frac{1}{4!}\frac{\partial^4 f}{\partial x^4}\Big|_{x=0}(x)^4 &= +\frac{1}{4!}e^{ix}\Big|_{x=0} x^4 = +x^4/4!
\end{aligned}
\quad \text{(A.19)}
$$

giving:

$$f(x) = 1 + ix - \frac{1}{2!}x^2 - i\frac{1}{3!}x^3 + \frac{1}{4!}x^4 + \dots \quad \text{(A.20)}$$

Collecting Real and Imaginary parts:

$$Re(f(x)) = 1 - \frac{1}{2!}x^2 + \frac{1}{4!}x^4 + \dots \quad \text{(A.21)}$$

$$Im(f(x)) = \frac{1}{1!}x - \frac{1}{3!}x^3 + \frac{1}{5!}x^5 + \dots \quad \text{(A.22)}$$

Setting $x = \theta$ and equating with the decomposition of $e^{i\theta}$ into Real and Imaginary parts (see Appendix A.1.4) yields the expansions for $\cos(\theta)$ and $\sin(\theta)$:

$$e^{i\theta} = \cos(\theta) + i\sin\theta \quad \text{(A.23)}$$

where:

$$cos(\theta) = 1 - \frac{1}{2!}\theta^2 + \frac{1}{4!}\theta^4 + \ldots$$

$$sin\theta = \theta - \frac{1}{3!}\theta^3 + \frac{1}{5!}\theta^5 + \ldots \quad (A.24)$$

A.1.3.5   Approximations, ($\mathcal{O}(n)$) Calculations, and Significant Figures

The unremitting emphasis in high school physics courses on the "right" answer can inadvertently develop a discomfort with approximation. A benefit from starting with a real theory, in this case Special Relativity, and after expanding energy and momentum each in a Taylor series, working to second order in v/c to acquire Newtonian mechanics, is that it precisely defines the level of approximation. You will encounter the phrase "working to order n" ($\mathcal{O}(n)$), or working to second order ($\mathcal{O}(2)$) in some variable (e.g., v/c). Nature is complex, and measurements always have a finite precision.

A related common mistake is to quote a numerical value with more digits, *significant figures*, than is known. Though rampant in media, please don't.

A.1.4   Imaginary Numbers and the Unit Circle: Euler's Formula

Figure A.2 shows a graphical representation of Eq. A.22.

From Eq. A.23, for $\theta = \pi$ we get Euler's Equation[5] relating $e, \pi, i, and -1$:

$$e^{i\pi} = \cos\pi + i\sin\pi$$

$$e^{i\pi} = -1. \quad (A.25)$$

Figure A.2 also shows the graphical representation of Eq. A.25. The y-axis is the imaginary part of the complex number $e^{i\pi}$; the x-axis is the real part. The magnitude of the vector r is $r = \sqrt{cos^2 + sin^2} = 1$; the locus traced by all values of θ is the "unit circle."

Note that if a uniform motion around the unit circle is projected on the x- and y-axes, i.e., one only looks at the one-dimensional motions, both are simple harmonic motion, out of phase by $90°$.

A.1.5   Cartesian, Polar, Cylindrical, and Spherical Coordinates

The choice of coordinate system, whether Cartesian, Cylindrical, or Spherical[6] is arbitrary, but calculations are much simpler in a coordinate system that corresponds to the symmetries of the problem.

---

[5] As spare as it gets.

[6] There are other interesting ones, e.g., hyperbolic.

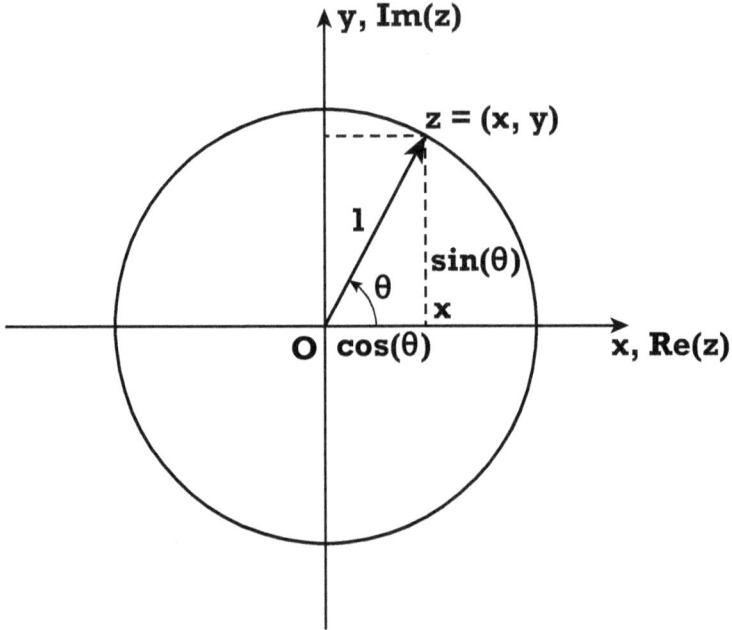

Figure A.2. The unit circle swept out by $z = e^{i\theta}$, where $z = x + iy$. The x and y coordinates are the Real and Imaginary parts of z, respectively.

A.1.5.1   *Unit Vectors*

A unit vector, denoted with a "hat" as in $\hat{x}$ and $\hat{\theta}$, is a vector of length one, i.e.,

$$\hat{A} \equiv \frac{\vec{A}}{|\vec{A}|}. \tag{A.26}$$

A vector $\vec{A}$ can thus be written as the product of the (scalar) magnitude and the unit vector:

$$\vec{r} = |\vec{r}|\hat{r}. \tag{A.27}$$

A.1.5.2   *Cartesian Coordinates*

We take the 3 unit vectors $\hat{x}, \hat{y}$, and $\hat{z}$ to form a right-handed orthonormal system. Generalizing the notation, we can write these as $\hat{x_1}, \hat{x_2}, \hat{x_3}$, and further generalizing, $\hat{x_i}, i = 1, 3$, where i is an index. Further:

- Orthonormal means the unit vectors have length one, $|\hat{x_i}|^2 = 1$, and the unit vectors are orthogonal, $\hat{x_i} \cdot \hat{x_j} = 0$ for $i \neq j$.

- Right-handed means $\hat{x} \times \hat{y} = \hat{z}$ and cyclic ($\hat{y} \times \hat{z} = \hat{x}$ and $\hat{z} \times \hat{x} = \hat{y}$);

- Cyclic means that x is followed by y is followed by z is followed by x. Anti-cyclic results from switching the order of two consecutive coordinates. In index notation, 123, 231, and 312 are cylic; 213, 321, and 132 are anticyclic.

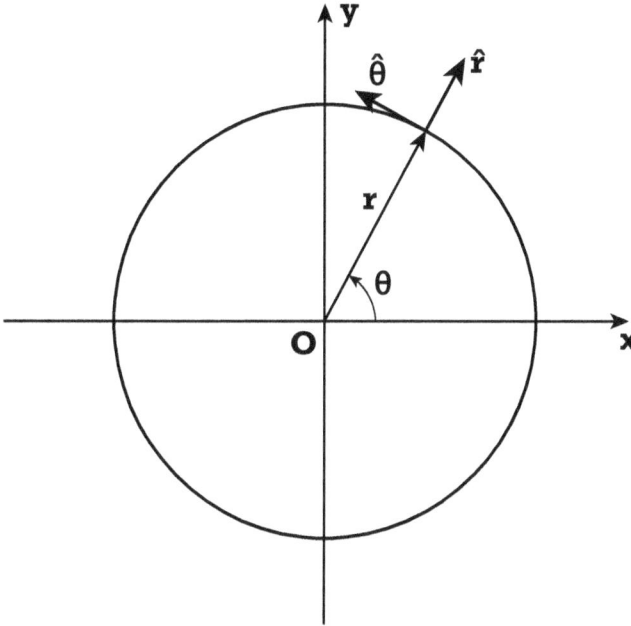

Figure A.3. Polar Coordinates. Our conventions are $\hat{x}$ and $\hat{y}$ form a right-handed system in the xy plane; the angle θ to the radius r is measured counter-clockwise from the positive x axis.

In tensor notation $\hat{x}_i \cdot \hat{x}_j = \delta_{ij}$, where δ_{ij} is the Kronecker delta and, $\delta_{ij} = 1$ if $i = j$, and $\delta_{ij} = 0$ otherwise.[7]

A.1.5.3 *Polar Coordinates*

Polar coordinates are defined in a 2-dimensional plane, here taken to be the xy plane. A point in the plane is specified by r, θ, where r is the (scalar) distance from the origin, and θ is the angle[8] from the x-axis in the counter-clockwise direction.[9]

The transformations from a 2D Cartesian coordinate system with unit vectors $(\hat{x}, \hat{y})$ to a polar coordinate system with unit vectors $(\hat{r}, \hat{\theta})$ are:

$$\hat{r} = \hat{x}\cos\theta + \hat{y}\sin\theta$$
$$\hat{\theta} = -\hat{x}\sin\theta + \hat{y}\cos\theta$$

(A.28)

where θ is measured counter-clockwise from the positive x axis. Note that the directions of the axes in polar coordinates are not fixed, but depend on θ.

---

[7] The 2×2 unit matrix.

[8] Many authors use ϕ instead of θ, reserving θ for spherical coordinates. It doesn't matter.

[9] Note that the coordinate system is multi-valued, as going $\theta + n\pi$, n an integer, refers to the same point in the space. One needs to be very careful with the inverse trigonometric functions to end up in the correct quadrant. See Ref [38] for a historical solution now used widely in the visual entertainment industry. Also note the similarity to Figure A.2.

The location of the off-diagonal minus sign, whether in the top row or the second row, changes depending on whether one is rotating the coordinate system (this convention) or the vectors in a fixed coordinate system.[10] I recommend *always* checking consistency with your convention, in this case that when $\theta = 0$, $\hat{r}$ lines up with positive $\hat{x}$ and $\hat{\theta}$ with $\hat{y}$, and when $\theta = 90°$, $\hat{r}$ lines up with positive $\hat{y}$ and $\hat{\theta}$ with $-\hat{x}$.

The inverse transformation in this convention is:

$$\hat{x} = \hat{r}\cos\theta - \hat{\theta}\sin\theta$$
$$\hat{y} = \hat{r}\sin\theta + \hat{\theta}\cos\theta. \tag{A.29}$$

A common but subtle misconception relates to the unit vector $\hat{\theta}$. Although the angle θ is dimensionless (it's a trigonometric function of a ratio of lengths), $\hat{\theta}$ is a vector with length 1, as is its partner $\hat{r}$.

A.1.5.4   *Cylindrical Coordinates*

Cylindrical coordinates are the 3-dimensional extension of polar coordinates, with the z-axis making a right-handed system with the polar xy plane. The coordinates of a point are r, θ, z. Note that $\hat{r} \times \hat{\theta} = \hat{z}$, i.e., the three unit vectors form a right-handed coordinate system rotated in the xy plane about the z-axis by θ.

Length, Area, and Volume: Only 2 of the 3 coordinates have units of length, as the angle is dimensionless. The distance-squared[11] to a point in the space from the origin is $l^2 = r^2 + z^2$.

A differential area on a cylinder around the z axis at radius r is represented as:

$$d^2A = (dr)(rd\theta). \tag{A.30}$$

A differential volume on a cylinder around the z axis at radius r is represented as:

$$d^3V = d^2Adz = (dr)(rd\theta)(dz). \tag{A.31}$$

A.1.5.5   *Spherical Coordinates*

The coordinates of a point $\vec{r}$ are r, θ, ϕ, where r is the distance to the origin, ϕ is the polar angle in the xy plane made by dropping a perpendicular from $\vec{r}$, as shown in Fig A.4.

Length, Area, and Volume Only 1 of the 3 coordinates has units of length, as angles are dimensionless. The distance-squared to a point in the space from the origin is $l^2 = r^2$.

---

[10] In that case (θ goes to $-\theta$, so $sin(\theta)$ goes to $-sin(\theta)$, and the minus sign moves from the second row to the first row.

[11] Square-root signs are a pain; feel free to write them in if you need to. I don't, and you won't.

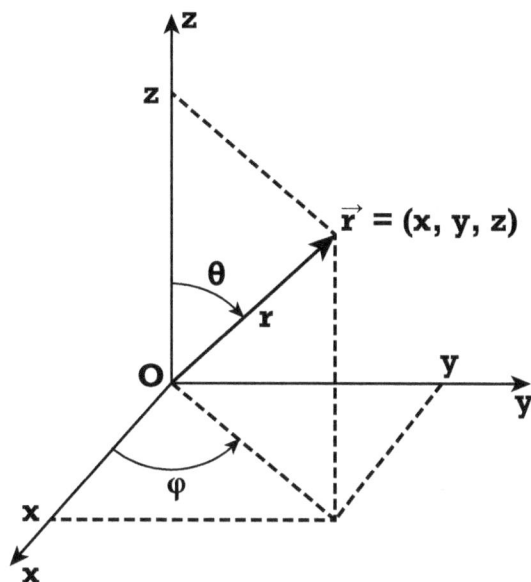

Figure A.4. Spherical Coordinates. Our conventions are that the z-axis is up, the polar angle θ is measured from the positive z-axis and the azimuthal angle ϕ is measured counter-clockwise from the positive x-axis. Note that $0 \leq \theta \leq \pi$ and $0 \leq \phi \leq 2\pi$.

A differential area on a sphere of radius r is represented as:

$$d^2A = (rd\cos(\theta))(rd\phi) \equiv r^2 \sin(\theta)d\theta d\phi \equiv r^2 d^2\Omega \tag{A.32}$$

where $d^2\Omega$ is a small patch[12] of solid angle $d^2\Omega \equiv (d\cos(\theta))(d\phi)$.
A differential volume is:

$$d^3V = r^2 dr d(\cos(\theta))d\phi. \tag{A.33}$$

A.1.6   Projection Operators

A.1.6.1   Projection Operators in Cartesian 3-Space

Consider the 3-vector $\vec{A} = (A_x, A_y, A_z) = A_x\hat{x} + A_y\hat{y} + A_z\hat{z}$. We can project out the x component by taking the scalar product with $\hat{x}$: $A_x = \hat{x} \cdot \vec{A}$, as $\hat{x}, \hat{y}$, and $\hat{z}$ form an orthonormal basis in 3-space.

Abstracting the operation, $\hat{x}\cdot$ (x - hat dot) is an operator that projects out the x-component of a vector in 3-space. Note that the convention is that the operator acts to the right.

A.1.6.2   Projection Operators in Quantum Mechanics: Recommended Reading

This is beyond the scope of the present course, but is not hard and will be helpful soon. As noted in the Introduction, Classical Mechanics is a limiting case of

---

[12] Corresponding to the area on the surface of the unit sphere.

Quantum Mechanics. While the context and meaning of projection operators is different in CM and QM, having the concept become natural early is important. Chapter 1 of Shankar is highly recommended [26].

A.1.7   Differential Operators, and the Vector Differential Operator Del in Cartesian Coordinates

The vector differential operator Del is defined as:

$$\nabla \equiv \hat{x}\frac{\partial}{\partial x} + \hat{y}\frac{\partial}{\partial y} + \hat{z}\frac{\partial}{\partial z}. \tag{A.34}$$

Because ∇ is a vector operator, it can operate on a scalar function to produce a vector function as above (Gradient), and also can operate on a vector function in a scalar product (Divergence), or in a vector product (Curl).[13] By convention, these differential operators operate to their right, and so $\nabla f \neq f\nabla$.

A.1.7.1   *The Gradient of a Scalar Function*

Consider a function in 3-dimensions, $f(x,y,z)$. The slope in the x direction $\hat{x}$ at point (x,y,z) is given by:

$$\frac{\partial f(x,y,z)}{\partial x}. \tag{A.35}$$

The slopes in the y and z directions have the same form, so the vector for the overall slope, i.e., the direction that the function increases most rapidly, is:

$$\nabla f(x,y,z) = \frac{\partial f(x,y,z)}{\partial x}\hat{x} + \frac{\partial f(x,y,z)}{\partial y}\hat{y} + \frac{\partial f(x,y,z)}{\partial z}\hat{z}. \tag{A.36}$$

Abstract the operation by defining the operator Del

$$\nabla \equiv \hat{x}\frac{\partial}{\partial x} + \hat{y}\frac{\partial}{\partial y} + \hat{z}\frac{\partial}{\partial z}. \tag{A.37}$$

The slope of the scalar function is given by the vector ∇f. Note that the slope, or gradient, points in the direction of maximum increasing f. Forces derived from conservative potentials point in the opposite direction, i.e., $\vec{F} = -\nabla V$.

A.1.7.2   *The Curl of a Vector Function*

The curl of a vector function $\vec{F} = F_x\hat{x} + F_y\hat{y} + F_z\hat{z}$ is a vector function:

---

[13] For a masterful summary of generalized coordinate systems see Ref. [39].

$$\nabla \times \vec{F} \equiv (\frac{\partial F_z}{\partial y} - \frac{\partial F_y}{\partial z})\hat{x} + (\frac{\partial F_x}{\partial z} - \frac{\partial F_z}{\partial x})\hat{y} + (\frac{\partial F_y}{\partial x} - \frac{\partial F_x}{\partial y})\hat{z}. \tag{A.38}$$

In index notation, the curl of $\vec{F}$ is given by:

$$(\nabla \times \vec{F})_i \equiv \nabla_j F_k - \nabla_k F_j \tag{A.39}$$

where the indices i, j, k run from 1 to 3 in cyclic order.[14]

A.1.7.3   *The Divergence of a Vector Function*

The divergence of a vector function $\vec{F}(x.y, z)$ is given by the scalar function:

$$\nabla \cdot \vec{F}(x, y, z) \equiv \frac{\partial F_x}{\partial x} + \frac{\partial F_y}{\partial y} + \frac{\partial F_z}{\partial z} \tag{A.40}$$

In index notation,

$$\nabla \cdot \vec{F} = \sum_{i=1,3} \frac{\partial F_i}{\partial x_i} \tag{A.41}$$

In index notation with the Einstein summation convention:

$$\nabla \cdot \vec{F} = \frac{\partial F_i}{\partial x_i} \tag{A.42}$$

A.1.7.4   *The Laplacian*

The scalar operator $\nabla \cdot \nabla$ is called the Laplacian.

$$\nabla \cdot \nabla = \frac{\partial^2}{\partial x^2} + \frac{\partial^2}{\partial y^2} + \frac{\partial^2}{\partial z^2}. \tag{A.43}$$

In index notation,

$$\nabla \cdot \nabla = \sum_{i=1,3} \frac{\partial^2}{\partial x_i^2}. \tag{A.44}$$

A.1.8   The Levi-Civita and Kronecker Delta Symbols

Tensor notation is succinct. Here we use the scalar and and cross products of two 3-vectors as examples. We use the Einstein summation convention, and leave the verification of the two formulae to the reader.

---

[14] Cyclic order is any 3 consecutive numbers in the sequence 123123123 ad infinitum, i.e, 123, 231, or 312. Anticyclic occurs when 2 integers in a cyclic triplet are switched: 132, 213, or 321. Also see the discussion of the Levi-Civita tensor in Section A.1.8.

1. **The Scalar (Dot) Product of 2 vectors in tensor notation**
   We can express the dot product of two vectors, $\vec{A} \cdot \vec{B}$ in tensor notation.

   The Kronecker delta symbol is written as δ_{ij}, where each of the 2 indices i,j may take on the values 1,2, or 3. The symbol δ_{ij} depends on the values of i and j as follows:

   $$\delta_{ij} = 1 \text{ if } i = j$$
   $$\delta_{ij} = 0 \text{ if } i \neq j.$$

   By writing it out explicitly one can show that

   $$\sum_{i=1,3}\sum_{j=1,3} \delta_{ij}A_iB_j = \delta_{ij}A_iB_j = \vec{A} \cdot \vec{B}. \tag{A.45}$$

2. **The Vector (Cross) Product of 2 vectors in tensor notation**

   The Levi-Civita symbol is written ϵ_{ijk}, where the 3 indices i,j,k take the values 1,2, or 3.

   The symbol ϵ_{ijk} depends on the values of i,j,k as follows:

   $\epsilon_{ijk} = +1$ if i,j,k are all different and are in cyclic order (123, 231, or 312);

   $\epsilon_{ijk} = -1$ if i,j,k are all different and are in anti-cyclic order (132, 213, or 321);

   $\epsilon_{ijk} = 0$ if any of i,j,k are the same (111,112,113,121,122,211, etc.).

   Consider the cross product $\vec{C} = \vec{A} \times \vec{B}$. Writing it out explicitly (many terms, all but two are zero), you can show that $C_i = \sum_{j=1,3}\sum_{k=1,3} \epsilon_{ijk}A_jB_k = \epsilon_{ijk}A_jB_k$ by writing out and evaluating all the terms.

A.1.9   The Einstein Summation Convention

We use the Einstein Summation Convention for matrix multiplication:

A repeated index is summed over, 1 to 3 in 3-space, 0 to 3 in 4-space.

For example, the scalar (dot) product of two 3-vectors $\vec{A}$ and $\vec{B}$ is succinctly written with the Kronecker delta as:

$$\vec{A} \cdot \vec{B} = \delta_{ij}A_iB_j. \tag{A.46}$$

The vector (cross) product of two 3-vectors $\vec{A}$ and $\vec{B}$ is succinctly written with Einstein summation as 3 equations with the Levi-Civita symbol as:

$$(\vec{A} \times \vec{B})_i = \epsilon_{ijk}A_jB_k. \tag{A.47}$$

RECOMMENDED READING

This is an eclectic list. For those who collect tangible books, used copies of many of these books are available at low prices from Alibris and other used book web sites.

Supplementary Texts

These are special books for the lay reader with a direct bearing on the course. While the course by design focuses on calculational tools and good practice, these texts provide a broader intellectual and historical background. They are special in the knowledge and experience of the authors, and in the depth and elegance of the exposition. I encourage using one or more of them for targeted reading week-by-week.

1. Robert N. Cahn and Chris Quigg, *Grace In All Simplicity: Beauty, Truth, and Wonders on the Path to the Higgs Boson and New Laws of Nature*; Pegasus Books; ISBN 978-1-63936-481-7 (2023).

   A remarkable book with a very personal and intelligent view of what we know about matter and how we know it. In addition to a deeply researched early history of electric charge, for example, the authors were deeply involved in modern particle physics. The descriptions of the discoveries of the charm quark and the Higgs boson are uniquely detailed and readable.

2. Frank Wilczek, *Fundamentals: Ten Keys to Reality*: Penguin Press; (2021); ISBN 9780735223790.

   Wilczek has written a highly readable but very deep book on our perceived Universe. To quote from the Preface, *"This is a book about fundamental lessons we can learn from the study of the physical world.... To me, those fundamental lessons include much more than bare facts about how the physical world works."* His book bears directly on issues that are foundational to a "radical revision" on how we teach introductory physics. The book is very special; I had not seen anything like this before.

Recommended Textbooks on Classical Mechanics

I recommend systematically exploring these. Each is special in its own way.

1. Jerry B. Marion, *Classical Dynamics of Particles and Systems*.
   A 2nd- or 3rd-year level text. See if you can find an early edition (Marion alone). The chapter on rigid body rotations and the Euler equations for stability is particularly good.

2. David Morin, *Introduction to Classical Mechanics*.
   Clear and deep—excellent. Chapter 7 is a particularly good source for Central Force Motion.

3. Cornelius Lanczos, *The Variational Principles of Mechanics*.
   A beautiful book.

4. Herbert Goldstein, *Classical Mechanics*.
   Too hard for an introductory course, but clear and definitive. Try it.[1]

5. M. E. Rose, *Elementary Theory of Angular Momentum*.
   Much too hard for a 1st-year course. Highly recommended.

6. M. J. Crowe, *A History of Vector Analysis*.
   Not essential for the course per se, but may be very helpful in providing a deeper understanding of the language we use. (It's also inspiring to learn about Grassman, Heaviside, and other remarkable autodidacts.)

7. R. Feynman, R. Leighton, and M. Sands, *The Feynman Lectures on Physics, Vol. I: The New Millennium Edition: Mainly Mechanics, Radiation, and Heat*.
   Feynman's lectures are deceptively sophisticated and I don't find them easy, but it's Feynman, and so worth reading.
   Available at https://www.feynmanlectures.caltech.edu/.

Books: Culture, History, and Humor

Physics is a culture replete with invention, humor, and occasionally, self-deprecation. This is a short personal list, chosen among many.

1. Owen Gingerich, *The Book Nobody Read*.
   A remarkable personal account of detective work with the manuscripts of Copernicus. It's also thoughtful and very knowledgeable on Kepler's Laws and ellipses.

2. Dava Sobel, *Galileo's Daughter*.
   A well-written history of Galileo.

---

[1] The footnote on p.161 is one of the classic physics footnotes; another, non-PC, is in Goldberger and Watson.

3. Joshua Gilder and Anne-Lee Gilder, *Heavenly Intrigue: Johannes Kepler, Tycho Brahe, and the Murder Behind One of History's Greatest Scientific Discoveries*. The remarkable story of Kepler and how he finally succeeded in getting Tycho Brahe's log-books of history-changing precise astronomical data. [19]

4. Graham Farmelo, *The Strangest Man*. An excellent biography of Paul Adrian Maurice Dirac. Highly recommended.

5. Nancy Thorndike Greenspan, *The End of the Certain World*. An excellent biography of Max Born. The Göttingen mathematicians Hilbert, Klein, and Minkowski were highly influential. This was also the environment in which Emmy Noether did her remarkable foundational work on invariance and conserved quantities with which we open the text [13].

6. Leo Szilard, *Voice of the Dolphins*. A witty, pointed, and (unfortunately) prescient science fiction by one of the most influential scientists of the 20th century. Also, look up his 10 Commandments.

7. William Lanouette, *Genius in the Shadows*. An excellent biography of Leo Szilard. Szilard's accomplishments and deep concerns should be known to all physicists.

8. Richard P. Feynman and Ralph Leighton, *Surely You're Joking, Mr. Feynman!* This is a little appalling (trigger warning) in places, but highly recommended.

9. R. V. Jones, *The Wizard War*. The development of radar, imaginative responses to German technical and strategic wartime developments, and the Theory of Spoof, in the context of the Battle of Britain. Highly recommended.

10. Otto Frisch, *What Little I Remember*. On a walk through the woods, Lisa Meitner and Otto Frisch, her nephew, figured out that nuclear fission had been discovered. Meitner's not being honored with Hahn and Strassmann is yet another blot on the Nobel Prize. A lovely little book.

11. Chanda Prescod-Weinstein, *The Disordered Cosmos: A Journey into Dark Matter, Spacetime, and Dreams Deferred*. A bold, important, and personal book that bears on the context of this text.

12. Robert N. Cahn and Chris Quigg, *Grace In All Simplicity*. (See Supplementary Texts.)

13. Frank Wilczek, *Fundamentals*. (See Supplementary Texts.)

Historical Context: Original Papers

The text is purposefully spare, providing what is needed operationally and mathematically, but without a lot of explanatory prose. The first three chapters, for example, take the student from the heuristic "Gedanken" experiments of Einstein to being able to calculate the momentum of the b-quarks in Higgs decay at the Large Hadron Collider. To calculate the momentum of the b-quarks, the sole products of the decay, all one needs to know are the masses of the Higgs and the b-quark. It takes only a few lines. It is a wonderfully succinct language.

However, the history can be fascinating. In addition to the books listed in *Culture, History, and Humor*, the following list provides a start on historical context through translations of some original manuscripts.

1. Galileo Galilei, *Dialogues Concerning Two New Sciences*;
   Legare Street Press; ISBN-13: 9781015599710 (2022)

2. S. Chandrasekhar, *Newton's Principia for the Common Reader*; Clarendon Press; ISBN-13: 9780198517443 (1995).
   Chandrasekhar was fascinated by Newton.

3. A. Einstein and H. Minkowski, *The Principle of Relativity; Original Papers by A. Einstein and H. Minkowski*;
   Alpha Edition (2020) ISBN-13: 9789354007835.
   (Be careful: my ordered copy was poorly printed.)

4. H. A. Lorentz, *The Principle of Relativity: A Collection of Original Memoirs on the Special and General Theory of Relativity*; Hassell Street Press (2021); ISBN-13: 9781014770127.

5. R. Cahn and G. Goldhaber, *The Experimental Foundations of Particle Physics*; Cambridge University Press (2009); ISBN 978-0-521-52147-5.
   A compendium of original papers starting with the discoveries of the neutron, positron, muon, and pion and extending through much of particle physics. Each chapter has a clear introduction, exercises, and further reading, and additional references. An excellent source on the evolution of the detector technology that allows us to see so deeply into this world.

6. K. Kuehn, *A Student's Guide Through the Great Physics Texts: Volume II: Space, Time and Motion*; Springer (2016); ISBN-13:9781493943692.
   A 4-volume text replete with translations of original papers, exercises, and vocabulary.

Skills and Guidelines

1. Strunk and White, *The Elements of Style*.
   "The little book" of essential writing skills for physicists. You may know E.B. White from *Charlotte's Web* and *Stuart Little*.

2. Egil "Bud" Krogh and Matthew Krogh, *Integrity: Good People, Bad Choices, and Life Lessons from the White House*; Perseus Books.
   A lesson on ethics by a highly-ethical but very young aide to Richard Nixon. Relevant to the requirement that you write out the problem sets yourself after working through them with your study group. The role of Elvis Presley is an added bonus.

GLOSSARIES

Glossary for the Text

This glossary is for terms in the text. The definitions have purposely been kept to one line. Because relativistic kinematics is largely used in particle physics, astrophysics, and cosmology, the problems sets have a separate glossary.

Atwood's 1st law of CM.	If asked to work in a non-inertial frame, just say **"NO."**
conserved quantity.	In a given frame, the total is constant (e.g., momentum, electric charge).
dimensional analysis.	Shortcut calculational method requiring dimensional consistency.
frame of reference.	The coordinate system in which position, time, and hence motion are measured.
inertial frame.	A frame in which an isolated particle at rest stays at rest.
natural and SI Units.	The space/time conversion factor $c \equiv 1$ (NU); $c \approx 299,792,458$ m/s (SI units).
Noether's theorem.	The relation between invariance and conserved quantities (I'm told).
space.	We can meet again **here**. See Wilczek.
space-time.	In our convention, the 4D space $(ct, \vec{x})$ with metric (1, -1, -1, -1).
speed of light.	The speed of massless particles; the maximum speed for everything else.
the Principle of SR.	The Laws of Physics Are the Same in All Inertial Frames.
time.	We will address this **later**. See Wilczek in Recommended Reading.
translation invariance.	The Laws of Physics Are the Same at All Times and All Places.
units of time.	second (s), msec (ms) , μsec (μs), nsec (ns), psec (ps), fsec (fs), asec (as).
units of space.	SI: m, cm, mm, μm; English: ft, in, miles.

Glossary for Problem Sets 1 and 3

These problem sets have relativistic exercises that involve contemporary ongoing research. Remarkably, everything one needs to know to solve each problem is contained in its statement; one doesn't need to know what the names mean or what physical reality they represent. For example, to solve the kinematics of the decay of the Higgs boson into two b-quarks, all one needs to know is that the mass of the b-quark and its anti-particle the b-bar are equal and the Higgs mass is more than twice the b-quark mass. The rest of the problem follows directly from conservation of energy and momentum, and the Master Relation $E^2 = p^2 + m^2$.[1] Similarly, the termination of the energy spectrum of the highest cosmic rays is determined by the collisions of high-energy protons with the photons that make up the Cosmic Microwave Background (CMB). However, it is unnerving to "get such a large return from such a small investment of fact" (Mark Twain again), and so this Glossary provides only thumb-nail definitions.

Note that for brevity we use Particle to mean sub-atomic particle, for example, protons and electrons; or quarks and gluons. Also, note that the proton is not strictly an elementary particle, as now we know that it is a composite of quarks and gluons about 3×10^{-24} seconds across.

Auger Observatory	Proposed by Jim Cronin and Alan Watson in 1992 to measure ultra-high energy cosmic rays. The detector is located in Mendoza, Argentina and comprises a ground array covering $\approx 3000\ km^2$ and fluorescence detectors to measure air showers.
boson.	Particle with integer value of spin.
Delta (Δ).	The lowest energy excitation of the quarks and gluons in the proton, with a mass of 1238 MeV (the ground state, which is the proton, has a mass of 938 MeV).
electron.	The lightest charged lepton; mass $m_e \approx 0.51$ MeV. See Cahn and Goldhaber.
fermion.	Particle with half-integer value of spin.
gluon.	The force carrier of the Strong interaction. Gluons carry two distinct charges of the Strong interaction, and like quarks are unobservable.
GZK cutoff.	The Earth-frame energy at which an ultra-high energy cosmic ray proton colliding with a 3-degree (Kelvin) photon in the cosmic microwave background has enough energy in the center-of-mass to make a Delta. This removes the proton from the spectrum, "cutting it off."
hadron.	Fermion with Strong, Electromagnetic and Weak interactions.
Higgs boson.	A massive neutral boson invented to provide mass to particles, and discovered in 2012. See Cahn and Quigg.
lepton.	Fermion with Electromagnetic and Weak interactions but not Strong.
momentum 4-vector.	$p^\mu = (E, c\vec{p})$.
more Higgs bosons?	I think it unlikely that the fermion masses have a rich structure due to only one lonely boson (*the* Higgs). I have bet that there are more to be discovered.
muon.	The next-to-lightest charged lepton; mass $m_\mu \approx 105$ MeV, $\approx 200 \times m_e$.
neutrino.	A very light (MeV-scale) neutral lepton. Each of the 3 known neutrinos is associated with its own charged lepton, e, μ or τ. Complex and fascinating. See Cahn and Quigg.

---

[1] For personal descriptions of the Higgs discovery, see *Grace in All Simplicity* by Cahn and Quigg in Recommended Reading.

photon.   A massless neutral boson; the quantum of the electro-magnetic field.

pion.   The π^+, π^-, π^0 are composite bosons that are the lightest bound state of quarks, with masses $\frac{1}{7}$ that of the proton.

quark.   An elementary fermion that makes up all hadrons. Quarks have non-integral values of electric charge, either $+\frac{2}{3}$ (up quarks) or $-\frac{1}{3}$ (down quarks).

spin.   Intrinsic angular momentum of a particle in units of $\hbar$.

Standard Model.   A marvelous *ad hoc* model of matter and forces with dozens of free parameters.

tau.   The heaviest charged lepton; mass $m_\tau \approx 1800$ MeV, $\approx 3500 \times m_e$ and $\approx 1.9 \times m_{proton}$.

top and bottom quarks.   The "top" quark, charge $+\frac{2}{3}$, has a mass of ≈ 173 GeV, comparable to a Tungsten nucleus. The "bottom" quark, charge $-\frac{1}{3}$, has a mass of ≈ 4 GeV.

$W^\pm$ and Z^0 Bosons.   The massive (80 and 90 GeV, respectively) force carriers of the Weak interaction. The Z^0 mixes with the photon in interactions in which no electric charge is transferred.

REFERENCES

[1] Photo Credit: Timothy Ferris, *Galaxies*; Sierra Club Books, (1980), p. 125, ISBN-139780871562739.

[2] Timothy Ferris, *Galaxies*; Sierra Club Books (1980); ISBN-13 9780871562739.

[3] E. P. Wigner, "The Unreasonable Effectiveness Of Mathematics In The Natural Sciences;" Richard Courant Lecture in mathematical sciences delivered at New York University, May 11, 1959; *Communications on Pure and Applied Mathematics* 13, no. 1 (1960): p. 1-14 ISSN: 0010-3640.

[4] Sheila Tobias *They're Not Dumb, They're Different: Stalking the Second Tier (Occasional Paper on Neglected Problems in Science Education)*; Science News Books (1994); ISBN-13 978-9991755663.

[5] See the sweet and foundational story *City Speed Limit* and its following gloss *The Professor's Lecture on Relativity which caused Mr Tompkin's dream* in G. Gamow, *Mr. Tompkins in Paperback*; Cambridge University Press (1940), illustrated by the author and John Hookham.

[6] Interesting topics beyond the scope of the course arise naturally, examples being the invariance and nature of electric charge, the equality of the limiting speed c among electromagnetic, gravitational, and massive particles, the equality of inertial and gravitational mass, and advanced mathematical techniques such as Stokes' Theorem. See *For Instructors*. During Covid we made the weekly discussion sessions remote (Zoom), and combined the individual TA discussion sessions into one common Zoom meeting. Each meeting had several "Bicycle Talks" each by a TA on interesting topics, or on topics for which the class had questions. The TAs enjoyed it, the students liked it, and attendence at the discussion sessions was very high, as opposed to the usual attendance. One of the TA's talks was ground-breaking.

[7] A. Einstein, *Relativity: The Special and General Theory*; Ockham Publishing (2022) ISBN-13: 9781839193613. The discussion of the Einstein Gedanken experiments owes a lot to a first edition of Robert March's *Physics for Poets*

that I used in teaching a Physical Sciences course for non-science majors. R. March, *Physics for Poets*; McGraw-Hill (1978); ISBN-13: 9780070402430.

[8] For a truly remarkable book on Newton, see S. Chandrasekhar, *Newton's Principia for the Common Reader*; Clarendon Press (1995); ISBN-13: 97801985174434.

[9] For a remarkable telling of the history and context of the measurements of the speed of light, see Dorothy Michelson Livingston, *The Master of Light: A Biography of Albert A. Michelson*; Charles Scribner's Son's (1973).

[10] D. H. Frisch and L. Wilets, "Development of the Maxwell-Lorentz Equations from Special Relativity and Gauss's Law;" *American Journal of Physics* 24, p. 574 (1956); https://doi.org/10.1119/1.1934322.

[11] Mark Twain, *Life on the Mississippi*; J. R. Osgood & Co.; (1883); Boston.

[12] The convention used here is that of J. D. Bjorken and S. D. Drell, *Relativistic Quantum Fields*; McGraw-Hill (1965). See the first section of their Appendix A.

[13] C. Quigg, Colloquium: "A Century of Noether's Theorem;" Fermilab-Pub-19-159-T (2019); https://arxiv.org/abs/1902.01989. Also see D. E. Rowe, and M. Koreuber, *Proving It Her Way: Emmy Noether, a Life in Mathematics*; Springer (2020); ISBN: 3030628108; ISBN-13: 9783030628109.

[14] L. Okun took on this misconception head-on(!); see, for example, L.B. Okun, "Mass versus relativistic and rest masses;" *American Journal of Physics* 77, 430 (2009); https://doi.org/10.1119/1.3056168.

[15] Also known as the Mozzi-Chasles' theorem, as it seems Mozzi derived and published it in 1763 while Chasles' work dates to 1830. The history seems complicated and unenlightening; I debated, but needed to call it by a name. The best I can do for a reference is: See https://en.wikipedia.org/wiki/Chasles_theorem_(kinematics).

[16] Jerry B. Marion amd Stephen T. Thornton, *Classical Dynamics of Particles and Systems*; Harcourt Brace (1995); ISBN 0-03-097302-3. The discussion of rigid bodies, including the stability of a freely rotating rigid body [34], is remarkable.

[17] A broader view of the history is a book waiting to be written.

[18] One can find the Newton's original Latin statement of his three Laws. For Kepler it is not so simple. For a discussion of Kepler's three laws, see Owen Gingerich, *The Book Nobody Read: Chasing the Revolutions of Nicolaus Copernicus*; Penguin Group (2005); ISBN-13: 9780143034766. The first 9 chapters are a recounting of Gingerich's detective work on Copernicus' publications in the mid-1500s. In Chapter 10 he turns to Kepler, and writes the following paragraph on Kepler's achievements:

He had to adjust the position of the Earth's orbit to make it work, and when he did, the periodic five-degree error in the Mars prediction just melted away. That was the biggest single correction that Kepler made in predicting the positions of the planets, and he doesn't get much credit for it because the astronomers who later selected three of Kepler's discoveries and numbered them as three laws (perhaps to match Newton's three laws) simply passed over this one as being too obvious.

[19] Joshua Gilder and Anne-Lee Guilder, *Heavenly Intrigue: Johannes Kepler, Tycho Brahe, and the Murder Behind One of History's Greatest Scientific Discoveries*; Doubleday (2004), ISBN 0-385-50844-1. Recommended reading.

[20] S. Chandrasekhar, *The Series Paintings of Claude Monet and the Landscape of General Relativity*; talk at the dedication ceremonies for the Inter-University Centre for Astronomny and Astrophysics, Pune, India; 1992; Special Collections, Regenstein Library, University of Chicago.

[21] P.A.M. Dirac, "Quantized Singularities in the Electromagnetic Field;" *Proceedings Royal Society A* 133 (1931) pp. 60–72, Includes a strong statement on the relationship of physics and Mathematics (large M).

[22] A.S. Eddington; see for example, *The Philosophy of Physical Science*; University of Michigan Press, (1958) ISBN-13 978-0472060207.

[23] H. Goldstein, *Classical Mechanics*; Addison-Wesley (1950); ISBN-13: 9780201025125. There are later editions; any will do. A classic advanced text.

[24] E. F. Taylor and J. A. Wheeler, *Spacetime Physics*; W. H. Freeman (1992); ISBN-13 9780716723271.

[25] The Wikipedia article on Rømer is excellent. https://en.wikipedia.org/wiki/Ole_Romer

[26] For an excellent introduction to the mathematics of linear vector spaces see Chapter 1 of R. Shankar, *Principles of Quantum Mechanics*; Plenum Publishing Corporation (1994); ISBN 0306403978, ISBN-13 9780306403972.

[27] J.J. Beatty, J. Matthews, and S.P. Wakely; https://pdg.lbl.gov/2019/reviews/rpp2019-rev-cosmic-rays.

[28] We call this the "Atomic Bomb Approximation." See L. Szilard, "Calling All Stars," in *Voice of the Dolphins*; Simon and Schuster (1961).

[29] C. Lanczos, The Variational Principles of Mechanics; Dover Publications (1986); ISBN 0486650677; ISBN-13 9780486650678.

[30] R. P. Feynman and A. R. Hibbs, *Quantum Mechanics and Path Integrals*; emended by Daniel F. Styer; Dover Publications (2010); ISBN-10 0-486-47722-3; ISBN-13 978-0-486-47722-0. I had tried to read this when it was first published. In emending it Styer corrected more than 879 errors.

[31] M. L. Rose, *Elementary Theory of Angular Momentum*; Dover Publications (2011); ISBN 0486684806; ISBN-13 9780486684802. Not elementary.

[32] The charge in the Figure is moving with $\gamma = 4$, so $\beta^2 = \frac{\gamma^2 - 1}{\gamma^2} = 15/16$, where β is the velocity in natural units and γ is the Lorentz factor derived in the first Gedanken experiment. Don't worry, we will get there.

[33] For an excellent history including a debunking of the myth of the Coriolis effect in the England-Germany Falklands naval battle in 1914 in WWI, see Christopher M. Graney, *Wide of the mark by 100 yards: Textbooks and the Falklands Coriolis myth* Commentary & Reviews DOI:10.1063/PT.6.3.20220202b 2 Feb 2022 https://physicstoday.scitation.org/do/10.1063/pt.6.3.20220202b/full/.

[34] To better understand our own intuition on how objects naturally want to, and do, move, see https://rotations.berkeley.edu/a-tumbling-t-handle-in-space/.

[35] *"The Pierre Auger Observatory was conceived by Jim Cronin and Alan Watson at the 1991 International Cosmic Ray Conferencee in Dublin to address the . . . nature of the highest-energy cosmic rays. It was clear to them that only a very large detector would have the exposure to collect enough events to answer the questions raised by a century of . . . optical telescopes."* Source: Timeline of the Pierre Auger Observatory; https://www.auger.org/observatory/timeline-observatory.

[36] K. Greisen, "End to the cosmic-ray spectrum?" *Phys. Rev. Lett.* 16, 748 (1966).

[37] G. T. Zatsepin and V. A. Kuzmin, "Upper limit of the spectrum of cosmic rays" *JETP Lett.* 4, 78 (1966).

[38] Andrew J. Hanson, *Visualizing Quaternions*, Elsevier (2006).

[39] My go-to text for such basics is the remarkably clear book by Paul Lorrain and Dale Corson *Electromagnetic Fields and Waves*; W. H. Freeman (1991); ISBN-13 978-0716703310. There are many second-hand editions available at Alibris; for example.

[40] M. J. Crowe, *A History of Vector Analysis*; Dover Publications (2013); ISBN 0486788776; 9780486788777. Highly recommended.

[41] Photo credit: https://solarsystem.nasa.gov/resources/2298/earths-moon-and-jupiters-moons/?categorȳplanets_jupiter

[42] See H. A. Lorentz, *Lectures on Theoretical Physics*; Jennings Press (2007); ISBN 101406729035; ISBN-13978.1406729030.

INDEX

GPSR Authorized Representative: Easy Access System Europe - Mustamäe tee
50, 10621 Tallinn, Estonia, gpsr.requests@easproject.com